T0309058

Molecular Beams: Science and Engineering

Molecular Beams: Science and Engineering

Edited by Josh Owen

STATES
ACADEMIC PRESS

www.statesacademicpress.com

States Academic Press,
109 South 5th Street,
Brooklyn, NY 11249, USA

Visit us on the World Wide Web at:
www.statesacademicpress.com

© States Academic Press, 2023

This book contains information obtained from authentic and highly regarded sources. Copyright for all individual chapters remain with the respective authors as indicated. All chapters are published with permission under the Creative Commons Attribution License or equivalent. A wide variety of references are listed. Permission and sources are indicated; for detailed attributions, please refer to the permissions page and list of contributors. Reasonable efforts have been made to publish reliable data and information, but the authors, editors and publisher cannot assume any responsibility for the validity of all materials or the consequences of their use.

ISBN: 978-1-63989-728-5

Trademark Notice: Registered trademark of products or corporate names are used only for explanation and identification without intent to infringe.

Cataloging-in-publication Data

Molecular beams : science and engineering / edited by Josh Owen.
    p. cm.
Includes bibliographical references and index.
ISBN 978-1-63989-728-5
1. Molecular beams. 2. Molecular dynamics. I. Owen, Josh.
QC173.4.M65 M65 2023
539--dc23

Table of Contents

Preface VII

Chapter 1 **Inducing Enantio-Sensitive Permanent**
Multipoles in Isotropic Samples with
Two-Color Fields 1
Andres F. Ordonez and Olga Smirnova

Chapter 2 **Manipulation and Control of Molecular**
Beams: The Development of the Stark-Decelerator 19
Gerard Meijer

Chapter 3 **The Precision Limits in a Single-Event Quantum**
Measurement of Electron Momentum and Position 33
H. Schmidt-Böcking, S. Eckart, H. J. Lüdde,
G. Gruber and T. Jahnke

Chapter 4 **High-Resolution Momentum Imaging — From Stern's**
Molecular Beam Method to the COLTRIMS
Reaction Microscope 55
T. Jahnke, V. Mergel, O. Jagutzki, A. Czasch,
K. Ullmann, R. Ali, V. Frohne, T. Weber,
L. P. Schmidt, S. Eckart, M. Schöffler, S. Schößler, S.
Voss, A. Landers, D. Fischer, M. Schulz, A. Dorn, L.
Spielberger, R. Moshammer, R. Olson, M. Prior, R.
Dörner, J. Ullrich, C. L. Cocke and H. Schmidt-Böcking

Chapter 5 **Liquid Micro Jet Studies of the Vacuum Surface of**
Water and of Chemical Solutions by
Molecular Beams and by Soft X-Ray Photoelectron
Spectroscopy 121
Manfred Faubel

Chapter 6 **Stern-Gerlach Interferometry with the Atom Chip** 154
Mark Keil, Shimon Machluf, Yair Margalit,
Zhifan Zhou, Omer Amit, Or Dobkowski, Yonathan
Japha, Samuel Moukouri, Daniel Rohrlich, Zina
Binstock, Yaniv Bar-Haim, Menachem Givon, David
Groswasser, Yigal Meir and Ron Folman

Chapter 7   **From Hot Beams to Trapped Ultracold
Molecules: Motivations, Methods and
Future Directions**..192
N. J. Fitch and M. R. Tarbutt

Permissions

List of Contributors

Index

Preface

A molecular beam is created when a gas expands from a higher pressure region to a lower pressure region through a small opening. It is used to directly measure uptake on a prepared surface as well as to provide information on the scattering of molecules from surfaces. These techniques have been widely employed to evaluate reactive and non-reactive uptake of radical and non-radical species on ice and water surfaces. The creation of the supersonic molecular beam, which is produced by a free jet expansion source, has significantly increased the versatility of the molecular beam technique. This book is a valuable compilation of topics, ranging from the basic to the most complex advancements in the science and engineering of molecular beams. It consists of contributions made by international experts. A number of latest researches have been included to keep the readers up-to-date with the global concepts in this area of study.

All of the data presented henceforth, was collaborated in the wake of recent advancements in the field. The aim of this book is to present the diversified developments from across the globe in a comprehensible manner. The opinions expressed in each chapter belong solely to the contributing authors. Their interpretations of the topics are the integral part of this book, which I have carefully compiled for a better understanding of the readers.

At the end, I would like to thank all those who dedicated their time and efforts for the successful completion of this book. I also wish to convey my gratitude towards my friends and family who supported me at every step.

Editor

Inducing Enantio-Sensitive Permanent Multipoles in Isotropic Samples with Two-Color Fields

Andres F. Ordonez and Olga Smirnova

Abstract We find that two-color fields can induce field-free permanent dipoles in initially isotropic samples of chiral molecules via resonant electronic excitation in a one-3ω-photon versus three-ω-photons scheme. These permanent dipoles are enantiosensitive and can be controlled via the relative phase between the two colors. When the two colors are linearly polarized perpendicular to each other, the interference between the two pathways induces excitation sensitive to the molecular handedness and orientation, leading to uniaxial orientation of the excited molecules and to an enantio-sensitive permanent dipole perpendicular to the polarization plane. We also find that although a corresponding one-2ω-photon versus two-ω-photons scheme cannot produce enantiosensitive permanent dipoles, it can produce enantiosensitive permanent quadrupoles that are also controllable through the two-color relative phase.

1 Introduction

Chirality (handedness) is the geometrical property that allows us to distinguish a left hand from a right hand. Like hands, many molecules have two possible versions which are non-superimposable mirror images of each other (opposite enantiomers). This "extra degree of freedom" stemming from the reduced symmetry (lack of improper symmetry axes) of chiral molecules leads to interesting behavior absent in achiral molecules [1–6] with profound implications for biology [7, 8]. Furthermore, since

A. F. Ordonez
Max-Born-Institut, Max-Born-Str. 2A, 12489 Berlin, Germany
e-mail: ordonez@mbi-berlin.de

O. Smirnova (✉)
Max-Born-Institut, Max-Born-Str. 2A, 12489 Berlin, Germany

Technische Universität Berlin, Straße des 17. Juni 135, 10623 Berlin, Germany
e-mail: smirnova@mbi-berlin.de

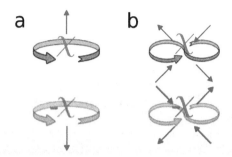

Fig. 1 Symmetry in ω and ω-2ω setups. **a** A circularly polarized field (circular arrow) interacts with an isotropic sample of chiral molecules (represented by χ) and produces a net photoelectron current (in general a vectorial signal) perpendicular to the polarization plane (arrow pointing up). The mirror reflection shows that the interaction of the same field with the opposite enantiomer (represented by $-\chi$) yields the opposite current. **b** A field with its fundamental and second harmonic linearly polarized perpendicular to each other (∞-like arrow) interacting with an isotropic chiral sample produces a quadrupolar photoelectron current (in general a quadrupolar signal). The mirror reflection shows that the interaction of the same field with the opposite enantiomer yields the opposite current. In both (**a**) and (**b**), the reversal of the signal when the polarization is changed follows from considering a rotation of 180° (not shown) of the full system, which changes the polarization but not the isotropic sample (see Figs. 2–4 in Ref. [14])

opposite enantiomers share fundamental properties like their mass and their energy spectrum, one must often rely precisely on this chiral behavior to tell opposite enantiomers apart—a task of immense practical importance in chemistry [9, 10].

An example of this chiral behavior is the phenomenon known as photoelectron circular dichroism (PECD) [4, 11–13], which consists in the generation of a net photoelectron current from an isotropic sample of chiral molecules irradiated by circularly polarized light [14–16]. This photoelectron current, which results from different amounts of photoelectrons being emitted in opposite directions, is directed along the normal to the polarization plane (because of the overall cylindrical symmetry) and changes sign when either the enantiomer or the circular polarization is reversed (see Fig. 1a). Importantly, PECD occurs within the electric-dipole approximation, which makes typical PECD signals orders of magnitude stronger than traditional enantiosensitive signals, such as circular dichroism (CD), which rely on interactions beyond the electric-dipole approximation [10, 17]. Furthermore, the electric-dipole approximation also rules out any influence of the wave vector of the incident light and hence of the momentum of the photons.

Given that: the molecules are randomly oriented in space, the electric field is circularly polarized, and the momentum of the photon does not play any role in PECD; it is only natural to wonder *why does a net current of photoelectrons perpendicular to the polarization plane occur?* From the point of view of symmetry, the question would be instead *what symmetry prevents this current from taking place in the case of achiral molecules?* The answer is simple: in the electric-dipole approximation[1] the

---

[1] Beyond the electric-dipole approximation the wave vector of the light breaks reflection symmetry.

system consisting of isotropic achiral molecules together with the circularly polarized electric field is symmetric with respect to reflection in the polarization plane[2] and therefore the current normal to the polarization plane must vanish. When achiral molecules are replaced by chiral molecules, this mirror symmetry is broken and the PECD current emerges [14].

While this symmetry analysis does not provide an answer in terms of the specific mechanism, the insight it provides applies to several other closely related effects occurring within the electric dipole approximation, which rely on electric field polarizations confined to a plane and yield enantiosensitive vectorial responses perpendicular to that plane [3, 14, 18–23]. For example, if the photon energy of the circularly polarized light is not enough to ionize the molecule, the lack of reflection symmetry due to the chiral molecules leads to oscillating bound currents normal to the polarization plane [14, 23]. In this case, the current results from the excitation of bound states and the associated oscillation of the expected value of the electric dipole operator. The enantiosensitivity is reflected in the phase of the oscillations, which are out of phase in opposite enantiomers.

Analogously, one may also expect that it should be possible to induce permanent electric dipoles (i.e. non-vanishing zero-frequency components of the expected value of the electric dipole operator) normal to the polarization plane and with opposite directions for opposite enantiomers. Indeed, such static electric dipoles have been investigated in the context of optical rectification [24–28], where two excited states close in energy are resonantly excited with monochromatic circularly polarized light. Very recently enantiosensitive static dipoles have also been studied in the context of molecular orientation induced by intense off-resonant light pulses [29–32]. Such light pulses excite rotational dynamics and cause orientation of one of the molecular axes that persists after the pulse is over. Here we show that field-free enantiosensitive permanent electric dipoles and the associated orientation can also be induced in the context of purely electronic excitation on ultrafast time-scales, without relying on rotational dynamics. We achieve this via interference of one- and three-photon excitation pathways.

Quite recently an extension of single-color PECD to two-color ω-2ω fields with orthogonal linear polarizations has been observed [33–35] (see Fig. 1b). As we discuss in Refs. [36, 37], this is an example of how molecular chirality can be reflected not only in scalar (e.g. CD) and vectorial observables (e.g. PECD), but also in higher-rank tensor observables. Here we show that two-color ω-2ω fields with linear polarizations perpendicular to each other can induce enantiosensitive permanent quadrupoles in samples of isotropic chiral molecules.

---

[2]Note that circularly polarized light is not chiral within the electric-dipole approximation, and therefore the chirality of the light itself does not play a role in PECD [14].

2 Exciting an Enantiosensitive Permanent Dipole

Consider the excitation scheme depicted in Fig. 2, where the interference of contributions from a one-3ω-photon pathway and a three-ω-photon pathway control the population of the state $|3\rangle$ of a chiral molecule. For simplicity, we first consider excitation via intermediate resonances in states $|1\rangle$ and $|2\rangle$. The presence of resonances in these states is not essential, as discussed later, but simplifies the analysis. The field is assumed to have the form

$$E^{\mathrm{L}}(t) = F(t)\left(E^{\mathrm{L}}_{\omega}e^{-i\omega t} + E^{\mathrm{L}}_{3\omega}e^{-3i\omega t}\right) + \text{c.c.}, \tag{1}$$

where $F(t)$ is a smooth envelope, E^{L}_{ω} and $E^{\mathrm{L}}_{3\omega}$ specify the polarizations and phases of each frequency, and the L and M superscripts indicate vectors and functions in the laboratory frame and in the molecular frame, respectively. For a given molecular orientation $\varrho \equiv \alpha\beta\gamma$, where $\alpha\beta\gamma$ are the Euler angles, the wave function after the interaction is

$$\Psi^{\mathrm{M}}(r^{\mathrm{M}}, \varrho) = \sum_{i=0}^{3} a_i(\varrho)\, e^{-i\omega_i t}\, \psi_i^{\mathrm{M}}(r^{\mathrm{M}}), \tag{2}$$

where $\psi_i^{\mathrm{M}}(r^{\mathrm{M}})$ is the coordinate representation of state $|i\rangle$ in the molecular frame. In the perturbative regime we have

$$a_3(\varrho) = A_3^{(1)}[d^{\mathrm{L}}_{3,0}(\varrho) \cdot E^{\mathrm{L}}_{3\omega}] + A_3^{(3)}[d^{\mathrm{L}}_{3,2}(\varrho) \cdot E^{\mathrm{L}}_{\omega}][d^{\mathrm{L}}_{2,1}(\varrho) \cdot E^{\mathrm{L}}_{\omega}][d^{\mathrm{L}}_{1,0}(\varrho) \cdot E^{\mathrm{L}}_{\omega}], \tag{3}$$

where $A_3^{(1)}$ and $A_3^{(3)}$ are first- and third-order coupling constants that depend on the detunings and the envelope (see Appendix). Analogous expressions apply for the other amplitudes a_i. The transition dipoles $d^{\mathrm{M}}_{i,j} \equiv \langle \psi_i^{\mathrm{M}}(r^{\mathrm{M}})|d^{\mathrm{M}}|\psi_j^{\mathrm{M}}(r^{\mathrm{M}})\rangle$ are fixed in the molecular frame and have been expressed in the laboratory frame using the rotation matrix $R(\varrho)$ according to $d^{\mathrm{L}}_{i,j}(\varrho) = R(\varrho)d^{\mathrm{M}}_{i,j}$.

Fig. 2 Excitation scheme used to produce an enantiosensitive permanent dipole in an isotropic sample of chiral molecules

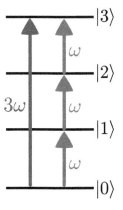

The expected value of the electric dipole operator in the molecular frame $\langle d^{\mathrm{M}}(\varrho)\rangle$ $\equiv \langle \Psi^{\mathrm{M}}(r^{\mathrm{M}}, \varrho)| d^{\mathrm{M}} |\Psi^{\mathrm{M}}(r^{\mathrm{M}}, \varrho)\rangle$ has a zero-frequency component of the form

$$\langle d^{\mathrm{M}}(\varrho)\rangle_{\omega=0} = \sum_{i=0}^{3} |a_i(\varrho)|^2 d_{i,i}^{\mathrm{M}}, \tag{4}$$

i.e. the permanent dipole for a given molecular orientation is the sum of the permanent dipoles of each state weighted by their orientation-dependent populations at the end of the pulse.

Transforming $\langle d^{\mathrm{M}}(\varrho)\rangle_{\omega=0}$ to the laboratory frame and averaging over all molecular orientations yields the permanent dipole

$$\langle d^{\mathrm{L}}\rangle_{\omega=0} \equiv \int d\varrho \langle d^{\mathrm{L}}(\varrho)\rangle_{\omega=0}. \tag{5}$$

The contribution of state $|3\rangle$ to this expression reads as[3]

$$\langle d_3^{\mathrm{L}}\rangle_{\omega=0} \equiv \int d\varrho \, |a_3(\varrho)|^2 d_{3,3}^{\mathrm{L}}(\varrho) = A_3^{(1)*} A_3^{(3)} \chi_3 Z^{\mathrm{L}} + \text{c.c.}, \tag{6}$$

where $\int d\varrho \equiv \int_0^{2\pi} d\alpha \int_0^{\pi} d\beta \int_0^{2\pi} d\gamma /8\pi^2$ is the integral over all molecular orientations and we defined

$$\chi_i \equiv \frac{1}{30} \left[(d_{2,1}^{\mathrm{M}} \cdot d_{1,0}^{\mathrm{M}}) d_{3,2}^{\mathrm{M}} + (d_{3,2}^{\mathrm{M}} \cdot d_{1,0}^{\mathrm{M}}) d_{2,1}^{\mathrm{M}} + (d_{3,2}^{\mathrm{M}} \cdot d_{2,1}^{\mathrm{M}}) d_{1,0}^{\mathrm{M}} \right] \cdot (d_{3,0}^{\mathrm{M}} \times d_{i,i}^{\mathrm{M}}), \tag{7}$$

$$Z^{\mathrm{L}} \equiv \left(E_\omega^{\mathrm{L}} \cdot E_\omega^{\mathrm{L}} \right) \left(E_\omega^{\mathrm{L}} \times E_{3\omega}^{\mathrm{L}*} \right). \tag{8}$$

χ_3 is a rotationally invariant molecular pseudoscalar, i.e. a molecular quantity independent of the molecular orientation. It has opposite signs for opposite enantiomers and vanishes for achiral molecules; χ_3 encodes the enantiosensitivity of $\langle d_3^{\mathrm{L}}\rangle_{\omega=0}$. Selection rules for χ_3 can be directly read off from Eq. (7). In particular, it vanishes if $d_{3,0}^{\mathrm{M}}$ and $d_{3,3}^{\mathrm{M}}$ are collinear. Z^{L} is a light pseudovector—it is a vector that depends only on the light's polarization and is invariant under the inversion operation; Z^{L} determines the direction of $\langle d_3^{\mathrm{L}}\rangle_{\omega=0}$. Selection rules for Z^{L} can be read off directly from Eq. (8). In particular, it vanishes if ω is circularly polarized ($E_\omega^{\mathrm{L}} \cdot E_\omega^{\mathrm{L}} = 0$) or if ω and 3ω are linearly polarized parallel to each other ($E_\omega^{\mathrm{L}} \times E_{3\omega}^{\mathrm{L}*} = 0$).

For example, if we choose ω and 3ω linearly polarized perpendicular to each other, say $E_\omega^{\mathrm{L}} = x^{\mathrm{L}}$ and $E_{3\omega}^{\mathrm{L}} = e^{-i\phi} y^{\mathrm{L}}$, with x^{L} and y^{L} the unitary vectors along each axis then

---

[3] We use Eq. (A16) in Ref. [14] for the interference term. The direct terms vanish because the possible non-zero field pseudovectors are purely imaginary, e.g. $(E_{3\omega}^* \times E_{3\omega})$, while the accompanying molecular pseudoscalars are real and the coupling coefficients appear within absolute values, see Ref. [38].

$$E^L(t) = 2F(t)\left[\cos(\omega t)\,x^L + \cos(3\omega t + \phi)\,y^L\right] \tag{9}$$

and we obtain

$$\langle d_3^L \rangle_{\omega=0} = 2\chi_3 \Re\left\{A_3^{(1)*} A_3^{(3)} e^{i\phi}\right\} z^L, \tag{10}$$

i.e., $\langle d_3^L \rangle_{\omega=0}$ is perpendicular to the polarization plane and its magnitude and sign can be controlled through the relative phase ϕ. Note that the relative phase of the coupling coefficients $A_3^{(1)}$ and $A_3^{(3)}$, which can be modified for example by changing the detunings, must also be taken into account.

The contributions from states $|1\rangle$, and $|2\rangle$ to the permanent dipole (5) have the same structure as Eq. (10), albeit with different coupling constants and molecular pseudoscalars χ_1 and χ_2, respectively [see Eq. (7)]. Since $|a_0|^2 = 1 - |a_1|^2 - |a_2|^2 - |a_3|^2$, the contribution from the ground state involves the coupling constants associated to $|1\rangle$, $|2\rangle$, and $|3\rangle$, and a molecular pseudoscalar χ_0 [see Eq. (7)]. Together, these contributions yield

$$\langle d^L \rangle_{\omega=0} = 2\left[(\chi_1 - \chi_0)\Re\left\{A_1^{(1)} A_1^{(3)*} e^{i\phi}\right\} + (\chi_2 - \chi_0)\Re\left\{A_2^{(2)'*} A_2^{(2)} e^{i\phi}\right\}\right.$$
$$\left. + (\chi_3 - \chi_0)\Re\left\{A_3^{(1)*} A_3^{(3)} e^{i\phi}\right\}\right] z^L \tag{11}$$

where $A_1^{(1)}$ and $A_1^{(3)}$ are the coupling coefficients for the transitions $|0\rangle \overset{\omega}{\to} |1\rangle$ and $|0\rangle \overset{3\omega}{\to} |3\rangle \overset{-\omega}{\to} |2\rangle \overset{-\omega}{\to} |1\rangle$, respectively; $A_2^{(2)}$ and $A_2^{(2)'}$ are the coupling coefficients for the transitions $|0\rangle \overset{\omega}{\to} |1\rangle \overset{\omega}{\to} |2\rangle$ and $|0\rangle \overset{3\omega}{\to} |3\rangle \overset{-\omega}{\to} |2\rangle$, respectively.

In the absence of the intermediate resonances through the states $|1\rangle$ and $|2\rangle$ the contribution from the third-order term in Eq. (3) turns into a sum over all intermediate states $|j\rangle$ and $|k\rangle$ weighted by a coefficient $A_{3;jk}^{(3)}$. The intermediate states retain no population at the end of the pulse and the permanent dipole takes the form

$$\langle d^L \rangle_{\omega=0} = 2\sum_{j,k}(\chi_{3;jk} - \chi_{0;jk})\Re\left\{A_3^{(1)*} A_{3;jk}^{(3)} Z^L\right\}$$
$$= 2F_0^4\,[2\pi\delta_\sigma(\Delta)]^2 \sum_{j,k}\frac{\chi_{3;jk} - \chi_{0;jk}}{(\omega_{k,0} - 2\omega_L)(\omega_{j,0} - \omega_L)}\Re\left\{Z^L\right\} \tag{12}$$

which is valid for arbitrary polarizations [see Eq. (8)]. Here $\chi_{i;jk}$ is given by Eq. (7) with the replacements $1 \to j$ and $2 \to k$. In the second equality we wrote the coupling constants explicitly, $\omega_{i,j} \equiv \omega_i - \omega_j$, $\Delta \equiv \omega_{3,0} - 3\omega_L$, and we took $\int_{-\infty}^{\infty} dt\, F(t)e^{i\omega t} \equiv 2\pi F_0 \delta_\sigma(\omega)$ with $\delta_\sigma(\omega)$ equal to the Dirac delta in the limit of infinitesimal σ.

2.1 A Simple Picture of the Mechanism Leading to the Enantiosensitive Permanent Dipole

The orientation averaging procedure we applied [38], although very powerful, is also rather formal. Below we demonstrate that the mechanism leading to the generation of the permanent dipole $\langle d^{\mathrm{L}} \rangle_{\omega=0}$ stems from the sensitivity of the excitation to the molecular orientation and handedness, which induces uniaxial and enantiosensitive orientation of the initially isotropic sample. We remark that the excitation induces *orientation* (↑) as opposed to just *alignment* (↕) and that this orientation is furthermore enantiosensitive.

Consider the interaction of the field (9) with a dummy molecule with $d^{\mathrm{M}}_{1,0}$ and $d^{\mathrm{M}}_{3,0}$ perpendicular to each other and $d^{\mathrm{M}}_{3,2} = d^{\mathrm{M}}_{2,1} = d^{\mathrm{M}}_{1,0}$. For simplicity we again assume that intermediate states are resonantly excited and that only state $|3\rangle$ has a non-zero permanent dipole. The population $P_3(\varrho) \equiv |a_3(\varrho)|^2$ of the excited state $|3\rangle$ reads [see Eq. (3)]

$$P_3(\varrho) = |A_3^{(1)}|^2 \mathcal{P}_{3\omega}(\varrho) + |A_3^{(3)}|^2 \mathcal{P}_{\omega}(\varrho) + 2\Re\left\{ A_3^{(1)*} A_3^{(3)} e^{i\phi} \right\} \mathcal{P}_{\omega,3\omega}(\varrho) \qquad (13)$$

where

$$\mathcal{P}_{\omega}(\varrho) \equiv \left[d^{\mathrm{L}}_{1,0}(\varrho) \cdot x^{\mathrm{L}} \right]^6, \quad \mathcal{P}_{3\omega}(\varrho) \equiv \left[d^{\mathrm{L}}_{3,0}(\varrho) \cdot y^{\mathrm{L}} \right]^2, \qquad (14)$$

$$\mathcal{P}_{\omega,3\omega}(\varrho) \equiv \left[d^{\mathrm{L}}_{3,0}(\varrho) \cdot y^{\mathrm{L}} \right] \left[d^{\mathrm{L}}_{1,0}(\varrho) \cdot x^{\mathrm{L}} \right]^3. \qquad (15)$$

$\mathcal{P}_{\omega}$ will select molecular orientations where $d^{\mathrm{L}}_{1,0}$ is aligned along the x^{L} axis. $\mathcal{P}_{3\omega}$ will select molecular orientations where $d^{\mathrm{L}}_{3,0}$ is aligned along the y^{L} axis. $\mathcal{P}_{\omega,3\omega}$ will select molecular orientations where $d^{\mathrm{L}}_{1,0}$ is aligned along the x^{L} axis *and* $d^{\mathrm{L}}_{3,0}$ is aligned along the y^{L} axis. These orientations are shown in Fig. 3b. While the direct terms $\mathcal{P}_{\omega}$ and $\mathcal{P}_{3\omega}$ do not distinguish between this subset of orientations $\{\varrho_i\}_{i=1}^4$, the interference term $\mathcal{P}_{\omega,3\omega}$ will be positive for orientations ϱ_1 and ϱ_3 and negative for orientations ϱ_2 and ϱ_4. This produces an imbalance between the number of excited molecules with orientations ϱ_1 and ϱ_3 and those with orientations ϱ_2 and ϱ_4. As can be seen in Fig. 3, this imbalance amounts to the molecular axis $d^{\mathrm{M}}_{1,0} \times d^{\mathrm{M}}_{3,0}$ being oriented. That is, the field (9) induces field-free uniaxial orientation of the molecular sample in the state $|3\rangle$. The emergence of a permanent dipole follows trivially, provided that $d^{\mathrm{M}}_{3,3}$ has a non-zero component along the oriented axis, i.e. as long as $d^{\mathrm{M}}_{3,3} \cdot (d^{\mathrm{M}}_{1,0} \times d^{\mathrm{M}}_{3,0}) \neq 0$. Note that, according to Eq. (7), this is in agreement with having $\chi_3 \neq 0$. If we consider the situation depicted in Fig. 3 now mirror reflected across the polarization plane, which is equivalent to swapping the enantiomer while leaving the field as it is, we immediately see that $d^{\mathrm{M}}_{3,3}$ and therefore also $\langle d^{\mathrm{L}}_3 \rangle_{\omega=0}$ point in the opposite direction, which explains the enantiosensitivity of $\langle d^{\mathrm{L}}_3 \rangle_{\omega=0}$.

Since the emergence of a permanent dipole $\langle d^{\mathrm{L}} \rangle_{\omega=0}$ relies on the molecules in the excited state $|3\rangle$ being oriented, we expect $\langle d^{\mathrm{L}} \rangle_{\omega=0}$ to survive for at least a few picoseconds before decaying due to molecular rotation. A decay of the dipole

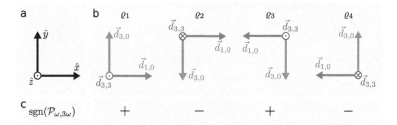

Fig. 3 Simplified analysis of the mechanism leading to an enantiosensitive permanent dipole for a field (9) and a dummy molecule with $\vec{d}_{1,0}$ and $\vec{d}_{3,0}$, perpendicular to each other and $\vec{d}_{3,2} = \vec{d}_{2,1} = \vec{d}_{1,0}$. Only the component of $\vec{d}_{3,3}$ perpendicular to the plane defined by $\vec{d}_{1,0}$ and $\vec{d}_{3,0}$ is shown. **a.** Laboratory frame. **b.** Molecular orientations with $\vec{d}_{1,0}$ aligned along x and $\vec{d}_{3,0}$ aligned along y. **c.** Sign of the interference term (15) for each molecular orientation. The interference distinguishes orientations ϱ_1 and ϱ_3 from orientations ϱ_2 and ϱ_4 and therefore causes the molecular axis $\vec{d}_{3,3}$ to become oriented. This leads to a non-vanishing permanent dipole

on the picosecond time-scale should lead to broadband THz emission [39] with an enantiosensitive phase. Furthermore, a quantum treatment of the rotational dynamics might reveal revivals of the molecular orientation (see e.g. Ref. [31]) .

3   Exciting an Enantiosensitive Permanent Quadrupole

Let us now consider the control scheme depicted in Fig. 4, where the interference of contributions from a one-2ω-photon pathway and a two-ω-photon pathway control the population of the state $|2\rangle$ of a chiral molecule. In this case the field reads as

$$\boldsymbol{E}^{\mathrm{L}}(t) = F(t)\left(\boldsymbol{E}_{\omega}^{\mathrm{L}}e^{-i\omega t} + \boldsymbol{E}_{2\omega}^{\mathrm{L}}e^{-2i\omega t}\right) + \text{c.c.} \qquad (16)$$

As in the previous section we begin assuming an intermediate resonance and then consider the case where the intermediate state is not resonant. The wave function reads as in Eq. (2) but with a sum up to $i = 2$,

$$a_2(\varrho) = A_2^{(1)}[\boldsymbol{d}_{2,0}^{\mathrm{L}}(\varrho) \cdot \boldsymbol{E}_{2\omega}^{\mathrm{L}}] + A_2^{(2)}[\boldsymbol{d}_{2,1}^{\mathrm{L}}(\varrho) \cdot \boldsymbol{E}_{\omega}^{\mathrm{L}}][\boldsymbol{d}_{1,0}^{\mathrm{L}}(\varrho) \cdot \boldsymbol{E}_{\omega}^{\mathrm{L}}]., \qquad (17)$$

Fig. 4 Excitation scheme
used to produce an
enantiosensitive permanent
quadrupole in an isotropic
sample of chiral molecules

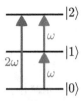

and an analogous expression for $a_1(\varrho)$. The expected value of the permanent electric quadrupole operator in the molecular frame $\langle Q_{p,q}^{M}(\varrho)\rangle \equiv \langle \Psi^{M}(r^{M},\varrho)| \, Q_{p,q}^{M} \, |\Psi^{M}(r^{M},\varrho)\rangle$, where $p,q = x,y,z$, will have a zero-frequency component of the form

$$\langle Q_{p,q}^{M}(\varrho)\rangle_{\omega=0} = \sum_{i=0}^{2} |a_i(\varrho)|^2 \langle Q_{p,q}^{M}\rangle_{i,i} \qquad (18)$$

where $\langle Q_{p,q}^{M}\rangle_{i,i} \equiv \langle \psi_i^{M}|Q_{q,p}^{M}|\psi_i^{M}\rangle$. Transforming $\langle Q_{p,q}^{M}(\varrho)\rangle_{\omega=0}$ to the laboratory frame and averaging over all molecular orientations yields the permanent quadrupole

$$\langle Q_{p,q}^{L}\rangle_{\omega=0} \equiv \int d\varrho \langle Q_{p,q}^{L}(\varrho)\rangle_{\omega=0}. \qquad (19)$$

The contribution of state $|2\rangle$ to this expression reads as (see Appendix)

$$\langle (Q_2^{L})_{p,q}\rangle_{\omega=0} \equiv \int d\varrho \, |a_2(\varrho)|^2 \langle Q_{p,q}^{L}(\varrho)\rangle_{2,2} \qquad (20)$$

$$= \langle (Q_2^{L})_{p,q}\rangle_{\omega=0}^{(\text{achiral})} + \left[A_2^{(1)*}A_2^{(2)} \chi_2' Z_{p,q}'^{L} + \text{c.c.} \right], \qquad (21)$$

where $\langle (Q_2^{L})_{p,q}\rangle_{\omega=0}^{(\text{achiral})}$ results from the diagonal terms in $|a_2(\varrho)|^2$ and is not enantiosensitive. χ_2' is a rotationally invariant molecular pseudoscalar (zero for achiral molecules) encoding the enantiosensitivity of $\langle (Q_2^{L})_{p,q}\rangle_{\omega=0}$ and defined according to

$$\chi_i' \equiv \frac{1}{30} \left\{ \left[(d_{1,0}^{M} \times d_{2,0}^{M}) \cdot (\langle Q^{M}\rangle_{i,i} d_{2,1}^{M}) \right] + \left[(d_{2,1}^{M} \times d_{2,0}^{M}) \cdot (\langle Q^{M}\rangle_{i,i} d_{1,0}^{M}) \right] \right\}, \qquad (22)$$

with $\langle Q^{M}\rangle_{i,i}$ a quadrupole matrix, i.e. $\langle Q^{M}\rangle_{i,i} d_{2,1}^{M}$ and $\langle Q^{M}\rangle_{i,i} d_{1,0}^{M}$ denote multiplications of a matrix and a vector. $Z_{p,q}'^{L}$ is a symmetric field pseudotensor of rank 2. It encodes the dependence of $\langle (Q_2^{L})_{p,q}\rangle_{\omega=0}$ on the field polarization according to

$$Z_{p,q}'^{L} \equiv \left(E_{\omega}^{L} \times E_{2\omega}^{L*}\right)_p \left(E_{\omega}^{L}\right)_q + \left(E_{\omega}^{L} \times E_{2\omega}^{L*}\right)_q \left(E_{\omega}^{L}\right)_p. \qquad (23)$$

This expression shows that all components of $Z_{p,q}'^{L}$ vanish if ω and 2ω are linearly polarized parallel to each other, or if ω and 2ω are circularly polarized and counter-rotating.

For example, if we take ω and 2ω linearly polarized perpendicular to each other, say $\boldsymbol{E}_\omega^L = \boldsymbol{x}^L$ and $\boldsymbol{E}_{2\omega}^L = e^{-i\phi}\boldsymbol{y}^L$, then

$$\boldsymbol{E}^L(t) = 2F(t)\left[\cos(\omega t)\,\boldsymbol{x}^L + \cos(2\omega t + \phi)\,\boldsymbol{y}^L\right], \tag{24}$$

and we obtain

$$\langle(Q_2^L)_{p,q}\rangle_{\omega=0} = \langle(Q_2^L)_{p,q}\rangle_{\omega=0}^{(\text{achiral})} + 2\chi_2\Re\left\{A_2^{(1)*}A_2^{(2)}e^{i\phi}\right\}\left(\delta_{p,z}\delta_{q,x} + \delta_{q,z}\delta_{p,x}\right). \tag{25}$$

Furthermore, one can show that for the field (24) the achiral terms vanish for $p \neq q$ (see Appendix) and therefore the enantiosensitive xz component reads as

$$\langle(Q_2^L)_{x,z}\rangle_{\omega=0} = 2\chi_2'\Re\left\{A_2^{(1)*}A_2^{(2)}e^{i\phi}\right\}, \tag{26}$$

i.e., it doesn't have an achiral background and can be controlled through the relative phase ϕ. The other non-diagonal components xy and yz vanish.

The contribution from state $|1\rangle$ to the permanent quadrupole (19) has the same structure as Eq. (26), although with different coupling constants and molecular pseudoscalar χ_1' [see Eq. (22)]. Since $|a_0|^2 = 1 - |a_1|^2 - |a_2|^2$, the contribution from the ground state involves the coupling constants associated to $|1\rangle$ and $|2\rangle$, and a molecular pseudoscalar χ_0' [see Eq. (22)]. Together, these contributions yield

$$\langle Q_{x,z}^L\rangle_{\omega=0} = 2\left[(\chi_1' - \chi_0')\Re\left\{A_1^{(1)}A_1^{(2)*}e^{i\phi}\right\} + (\chi_2' - \chi_0')\Re\left\{A_2^{(1)*}A_2^{(2)}e^{i\phi}\right\}\right], \tag{27}$$

where $A_1^{(1)}$ and $A_1^{(2)}$ are the coupling coefficients for the transitions $|0\rangle \overset{\omega}{\to} |1\rangle$ and $|0\rangle \overset{2\omega}{\to} |2\rangle \overset{-\omega}{\to} |1\rangle$, respectively. The other non-diagonal elements of the permanent quadrupole vanish and the diagonal terms are not enantiosensitive.

As in the previous section, in the absence of an intermediate resonance through the state $|1\rangle$, the contribution from the second-order term in Eq. (17) turns into a sum over all intermediate states $|j\rangle$. The intermediate states retain no population at the end of the pulse and the permanent quadrupole takes the form

$$\langle Q_{x,z}^L\rangle_{\omega=0} = 2\sum_j(\chi_{2;j}' - \chi_{0;j}')\Re\left\{A_2^{(1)*}A_{2;j}^{(2)}e^{i\phi}\right\}$$

$$= 2F_0^3\left[2\pi\delta_\sigma(\Delta)\right]^2\sum_j\frac{\chi_{2;j}' - \chi_{0;j}'}{\omega_{j,0} - \omega_L}\cos\phi, \tag{28}$$

where $\Delta \equiv \omega_{2,0} - 2\omega_L$, $\chi_{i;j}$ is given by Eq. (22) with the replacement $1 \to j$ and the other symbols were introduced as in Eq. (12).

A simplified analysis analogous to that presented in Fig. 3 is shown in Fig. 5 for the case of the field (24) interacting with a dummy molecule with $\boldsymbol{d}_{1,0}^M, \boldsymbol{d}_{2,0}^M$, and $\langle Q^M\rangle_{2,2}$ oriented as shown and with $\boldsymbol{d}_{2,1}^M = \boldsymbol{d}_{1,0}^M$. The population of state $|2\rangle$ is determined by

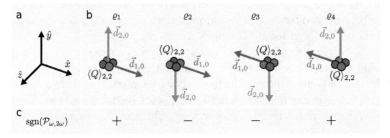

Fig. 5 Simplified analysis of the mechanism leading to an enantiosensitive permanent quadrupole for a field (24) and a dummy molecule with $d_{1,0}, d_{2,0}$, and $\langle Q \rangle_{2,2}$ oriented as shown with respect to each other and $d_{2,1} = d_{1,0}$. Blue and red balls stand for negative and positive charges. **a.** Laboratory frame. **b.** Molecular orientations with $d_{1,0}$ aligned along x and $d_{2,0}$ aligned along y. **c.** Sign of the interference term (29) for each molecular orientation. The interference causes the molecular axis $d_{2,0}$ to become oriented, which together with the alignment of $d_{1,0}$ along x explains the non-vanishing permanent quadrupole

the interference term

$$\mathcal{P}_{\omega,2\omega}(\varrho) \equiv [d_{2,0}(\varrho) \cdot y][d_{1,0}(\varrho) \cdot x]^2, \tag{29}$$

which is positive for orientations ϱ_1 and ϱ_4 and negative for orientations ϱ_2 and ϱ_3. This causes the molecular axis $d_{2,0}^{M}$ to become oriented. If $d_{2,2}^{M}$ has a non-zero component along $d_{2,0}^{M}$, then a permanent dipole emerges. However, this permanent dipole is contained in the polarization plane and will therefore not change upon reflection of the system across the polarization plane. Since this reflection is equivalent to a change of the enantiomer, the permanent dipole is not enantiosensitive. In contrast and as can be seen from Fig. 5, the imbalance between orientations ϱ_1 and ϱ_4 in comparison to orientations ϱ_2 and ϱ_3 is enough to produce an enantiosensitive permanent quadrupole that does change upon reflection in the polarization plane.

4   Conclusions

We have shown that permanent dipoles and quadrupoles can be induced in initially isotropic samples of chiral molecules using perturbative two-color fields that resonantly excite electronic transitions. These permanent multipoles are enantiosensitive and their sign can be controlled through the relative phase between the two colors. The mechanism leading to these permanent dipoles (or quadrupoles) stems from uniaxial orientation of the molecule, which occurs due to the selectivity of the excitation to the orientation of the molecule. Such orienting excitation can be accomplished using fields where the fundamental and its second (or third) harmonic are linearly polarized perpendicular to each other. The enantiosensitive permanent dipole is obtained via three-ω- versus one-3ω-photon interference. The enantiosensitive quadrupole is obtained via two-ω versus one-2ω interference. In the latter case, a permanent dipole can also be generated but it is not enantiosensitive. We expect these permanent multipoles to survive for at least a few picoseconds before decaying due to molecular rotation. Such picosecond variation of the multipoles should in principle lead to broadband THz emission with an enantiosensitive phase.

Although we focused on a mechanism relying on interference between two pathways, it is also possible to induce permanent dipoles via direct pathways by relying on transitions where the photon order matters. This can be achieved e.g. using the pulse sequence in Ref. [29].

Efficient generation of enantio-sensitive permanent dipoles and quadrupoles via orientation-sensitive excitations is possible in strong laser fields using efficient excitation of Rydberg states via the so-called Freeman resonances [40] in the regime when the pronderomotive potential is comparable to the laser frequency. Since Rydberg states have large polarizability, we expect significant contrast in the orientation of left and right enantiomers. Opposite orientation of left and right enantiomers and their respective induced permanent dipoles create opportunities for enantio separation using static electric fields.

Acknowledgements We gratefully acknowledge support from the DFG SPP 1840 "Quantum Dynamics in Tailored Intense Fields" within the project SM 292/5-1;

Appendix

Coupling coefficients $A_f^{(n)}$

Consider a Hamiltonian $H = H_0 + H'(t)$, where H_0 is the time-independent field-free Hamiltonian and $H'(t)$ can be treated as a perturbation. If at time $t = 0$ the system is in the state $|0\rangle$, the probability amplitude of finding the system in the state $|f\rangle$ at the time $t = T$ can be written as $a_f = a_f^{(1)} + a_f^{(2)} + \cdots$, where

$$a_f^{(N)} = \left(\frac{1}{i}\right)^N \int\limits_0^T dt_N \ldots \int\limits_0^{t_3} dt_2 \int\limits_0^{t_2} dt_1 \langle f | H_I'(t_N) \ldots H_I'(t_2) H_I'(t_1) | 0 \rangle \qquad (30)$$

and $H_I'(t) = e^{iH_0 t} H'(t) e^{-iH_0 t}$. In the electric dipole approximation we have $H' = -\boldsymbol{d} \cdot \boldsymbol{E}(t)$. For a field $\boldsymbol{E}(t) = F(t)\boldsymbol{E}_\omega e^{-i\omega t} + \text{c.c.}$, the contributions to $a_f^{(N)}$ from absorption of N photons yield

$$a_f^{(N)} = \sum_{j_1, j_2, \ldots, j_{N-1}} a_{f; j_1, j_2, \ldots, j_{N-1}}^{(N)}, \qquad (31)$$

where the sum is over the different quantum pathways through the intermediate states $|j_1\rangle, |j_2\rangle, ..., |j_{N-1}\rangle$. The amplitude of each pathway can be written as

$$a^{(N)}_{f;j_1,j_2...j_{N-1}} = A^{(N)}_{f;j_1,j_2,...,j_{N-1}}(\omega)\,(d_{f,j_{N-1}} \cdot E_\omega)...(d_{j_2,j_1} \cdot E_\omega)(d_{j_1,0} \cdot E_\omega), \qquad (32)$$

The coupling coefficient $A^{(N)}_{f;j_1,j_2,...,j_{N-1}}(\omega)$ carries the information about the frequency of the light, its envelope, and the detunings according to

$$A^{(N)}_{f;j_1,j_2,...,j_{N-1}}(\omega) = i^N \int_0^T dt_N F(t_N)\, e^{i(\omega_{f,j_{N-1}} - \omega)t_N} ...$$

$$\times \int_0^{t_3} dt_2 F(t_2)\, e^{i(\omega_{j_2,j_1} - \omega)t_2} \int_0^{t_2} dt_1 F(t_1)\, e^{i(\omega_{j_1,0} - \omega)t_1}, \qquad (33)$$

where $\omega_{ij} \equiv \omega_i - \omega_j$. Contributions to $a^{(N)}_f$ from pathways involving photon emissions require exchanging E_ω by E^*_ω in Eq. (32)[4] and ω by $-\omega$ in Eq. (33) in the corresponding transitions.

In the case of a resonant pathway the sum in Eq. (31) reduces to a single term, which is the assumption in several parts of the main text. There we write $A^{(N)}_f$ as a shorthand for $A^{(N)}_{f;j_1,j_2,...,j_{N-1}}$.

Orientation integrals required in Sect. 3

Replacing Eq. (17) in Eq. (20) we obtain

$$\langle Q^{L}_{p,q}\rangle_{\omega=0} = |A^{(1)}_2|^2 I^{(2\omega)}_{p,q} + |A^{(2)}_2|^2 I^{(\omega)}_{p,q} + \left[A^{(1)*}_2 A^{(2)}_2 I^{(\omega,2\omega)}_{p,q} + \text{c.c.}\right], \qquad (34)$$

where the integrals $I^{(2\omega)}_{p,q}$, $I^{(\omega)}_{p,q}$, and $I^{(\omega,2\omega)}_{p,q}$ are defined by

$$I^{(2\omega)}_{p,q} \equiv \int d\varrho\, \left|d^{L}_{2,0}(\varrho) \cdot E^{L}_{2\omega}\right|^2 \langle Q^{L}_{p,q}\rangle_{2,2}, \qquad (35)$$

$$I^{(\omega)}_{p,q} \equiv \int d\varrho\, \left|[d^{L}_{2,1}(\varrho) \cdot E^{L}_{\omega}][d^{L}_{1,0}(\varrho) \cdot E^{L}_{\omega}]\right|^2 \langle Q^{L}_{p,q}\rangle_{2,2}, \qquad (36)$$

$$I^{(\omega,2\omega)}_{p,q} \equiv \int d\varrho\, [d^{L}_{2,1}(\varrho) \cdot E^{L}_{\omega}][d^{L}_{1,0}(\varrho) \cdot E^{L}_{\omega}][d^{L}_{2,0}(\varrho) \cdot E^{L*}_{2\omega}]\langle Q^{L}_{p,q}\rangle_{2,2}. \qquad (37)$$

---

[4]If the transition dipoles are complex then one must also complex conjugate them. Here we assume they are real.

These integrals can be solved following the procedure in Ref. [38]. We will first solve $I_{p,q}^{(\omega,2\omega)}$ for arbitrary polarizations and then show that $I_{p,q}^{(2\omega)}$ and $I_{p,q}^{(\omega)}$ vanish when $p \neq q$, $\boldsymbol{E}_{\omega}^{\mathrm{L}} = \boldsymbol{x}$, and $\boldsymbol{E}_{2\omega}^{\mathrm{L}} = e^{-i\phi}\boldsymbol{y}^{\mathrm{L}}$.

$I_{p,q}^{(\omega,2\omega)}$

We are dealing with an integral of the form

$$I_{i_4 i_5} = \int \mathrm{d}\varrho \left(\boldsymbol{a}^{\mathrm{L}} \cdot \boldsymbol{B}^{\mathrm{L}}\right) \left(\boldsymbol{b}^{\mathrm{L}} \cdot \boldsymbol{B}^{\mathrm{L}}\right) \left(\boldsymbol{c}^{\mathrm{L}} \cdot \boldsymbol{C}^{\mathrm{L}}\right) Q_{i_4,i_5}^{\mathrm{L}}$$

$$= I_{i_1 i_2 i_3 i_4 i_5;\lambda_1\lambda_2\lambda_3\lambda_4\lambda_5}^{(5)} a_{\lambda_1}^{\mathrm{M}} b_{\lambda_2}^{\mathrm{M}} c_{\lambda_3}^{\mathrm{M}} Q_{\lambda_4,\lambda_5}^{\mathrm{M}} B_{i_1}^{\mathrm{L}} B_{i_2}^{\mathrm{L}} C_{i_3}^{\mathrm{L}}, \tag{38}$$

where

$$I_{i_1 i_2 i_3 i_4 i_5;\lambda_1\lambda_2\lambda_3\lambda_4\lambda_5}^{(5)} \equiv \int \mathrm{d}\varrho\, l_{i_1\lambda_1} l_{i_2\lambda_2} l_{i_3\lambda_3} l_{i_4\lambda_4} l_{i_5\lambda_5}, \tag{39}$$

$\boldsymbol{a}^{\mathrm{M}}$, $\boldsymbol{b}^{\mathrm{M}}$, and $\boldsymbol{c}^{\mathrm{M}}$ are arbitrary vectors fixed in the molecular frame, and Q_{i_4,i_5}^{M} is an arbitrary symmetric second-rank tensor fixed in the molecular frame. The transformation to the laboratory frame is given by $v_i^{\mathrm{L}}(\varrho) = l_{i\lambda}(\varrho) v_{\lambda}^{\mathrm{M}}$ for vectors and $Q_{i_1,i_2}^{\mathrm{L}}(\varrho) = l_{i_1\lambda_1}(\varrho) l_{i_2\lambda_2}(\varrho) Q_{\lambda_1,\lambda_2}^{\mathrm{M}}$ for the second-rank tensor, where $l_{i\lambda}(\varrho)$ is the matrix of direction cosines, we sum over repeated indices and use latin indices for components in the laboratory frame and greek indices for components in the molecular frame. $\boldsymbol{B}^{\mathrm{L}}$ and $\boldsymbol{C}^{\mathrm{L}}$ are arbitrary vectors fixed in the laboratory frame. Using Eq. (31) in Ref. [38] we obtain

$$I_{i_4 i_5} = \frac{1}{30}\left[\epsilon_{\lambda_1\lambda_3\lambda_4}\delta_{\lambda_2\lambda_5}\epsilon_{i_1 i_3 i_4}\delta_{i_2 i_5} + \epsilon_{\lambda_1\lambda_3\lambda_5}\delta_{\lambda_2\lambda_4}\epsilon_{i_1 i_3 i_5}\delta_{i_2 i_4} + \epsilon_{\lambda_2\lambda_3\lambda_4}\delta_{\lambda_1\lambda_5}\epsilon_{i_2 i_3 i_4}\delta_{i_1 i_5}\right.$$

$$\left. + \epsilon_{\lambda_2\lambda_3\lambda_5}\delta_{\lambda_1\lambda_4}\epsilon_{i_2 i_3 i_5}\delta_{i_1 i_4}\right] a_{\lambda_1}^{\mathrm{M}} b_{\lambda_2}^{\mathrm{M}} c_{\lambda_3}^{\mathrm{M}} Q_{\lambda_4,\lambda_5}^{\mathrm{M}} B_{i_1}^{\mathrm{L}} B_{i_2}^{\mathrm{L}} C_{i_3}^{\mathrm{L}} \tag{40}$$

where we used $\epsilon_{i_1 i_2 i_3} B_{i_2} B_{i_3} = \epsilon_{\lambda_1\lambda_2\lambda_3} Q_{\lambda_2\lambda_3} = 0$. The first term can be rewritten as

$$\epsilon_{\lambda_1\lambda_3\lambda_4}\delta_{\lambda_2\lambda_5}\epsilon_{i_1 i_3 i_4}\delta_{i_2 i_5} a_{\lambda_1}^{\mathrm{M}} b_{\lambda_2}^{\mathrm{M}} c_{\lambda_3}^{\mathrm{M}} Q_{\lambda_4,\lambda_5}^{\mathrm{M}} B_{i_1}^{\mathrm{L}} B_{i_2}^{\mathrm{L}} C_{i_3}^{\mathrm{L}}$$

$$= \left(\boldsymbol{a}^{\mathrm{M}} \times \boldsymbol{c}^{\mathrm{M}}\right)_{\lambda_4} Q_{\lambda_4,\lambda_5}^{\mathrm{M}} b_{\lambda_5}^{\mathrm{M}} \left(\boldsymbol{B}^{\mathrm{L}} \times \boldsymbol{C}^{\mathrm{L}}\right)_{i_4} B_{i_5}^{\mathrm{L}}$$

$$= \left[\left(\boldsymbol{a}^{\mathrm{M}} \times \boldsymbol{c}^{\mathrm{M}}\right) \cdot \left(\boldsymbol{Q}^{\mathrm{M}} \boldsymbol{b}^{\mathrm{M}}\right)\right] \left(\boldsymbol{B}^{\mathrm{L}} \times \boldsymbol{C}^{\mathrm{L}}\right)_{i_4} B_{i_5}^{\mathrm{L}}. \tag{41}$$

Analogous operations for the rest of the terms yield

$$I_{i_4 i_5} = \frac{1}{30} \left\{ \left[(\boldsymbol{a}^M \times \boldsymbol{c}^M) \cdot (Q^M \boldsymbol{b}^M) \right] + \left[(\boldsymbol{b}^M \times \boldsymbol{c}^M) \cdot (Q^M \boldsymbol{a}^M) \right] \right\}$$
$$\times \left\{ (\boldsymbol{B}^L \times \boldsymbol{C}^L)_{i_4} B_{i_5}^L + (\boldsymbol{B}^L \times \boldsymbol{C}^L)_{i_5} B_{i_4}^L \right\} \tag{42}$$

Performing the substitutions $\{a, b, c, Q\} \to \{d_{2,1}, d_{1,0}, d_{2,0}, \langle Q \rangle_{2,2}\}$, $\{B, C\} \to \{E_\omega, E_{2\omega}^*\}$, and $\{i_4, i_5\} \to \{p, q\}$ and using Eqs. (34) and (37) yields Eqs. (21)-(23).

$I_{p,q}^{(2\omega)}$

Assuming a linearly polarized $\boldsymbol{E}_{2\omega}$ we must deal with an integral of the form

$$I_{i_3 i_4} = \int d\varrho \, [\boldsymbol{a}^L \cdot \boldsymbol{B}^L][\boldsymbol{a}^L \cdot \boldsymbol{B}^L] Q_{i_3 i_4}^L$$
$$= I_{i_1 i_2 i_3 i_4 ; \lambda_1 \lambda_2 \lambda_3 \lambda_4}^{(4)} a_{\lambda_1}^M a_{\lambda_2}^M Q_{\lambda_3, \lambda_4}^M B_{i_1}^L B_{i_2}^L, \tag{43}$$

where

$$I_{i_1 i_2 i_3 i_4 ; \lambda_1 \lambda_2 \lambda_3 \lambda_4}^{(4)} \equiv \int d\varrho \, l_{i_1 \lambda_1} l_{i_2 \lambda_2} l_{i_3 \lambda_3} l_{i_4 \lambda_4}, \tag{44}$$

and we use the same notation as in the previous subsection. Using Eq. (19) in Ref. [38] we get

$$I_{i_3 i_4} = \boldsymbol{F}_{i_3 i_4}^{(4)} \cdot M^{(4)} \boldsymbol{G}_{i_3 i_4}^{(4)}, \tag{45}$$

where $\boldsymbol{F}_{i_3 i_4}^{(4)}$ is given by

$$\boldsymbol{F}_{i_3 i_4}^{(4)} = \begin{pmatrix} \delta_{i_1 i_2} \delta_{i_3 i_4} \\ \delta_{i_1 i_3} \delta_{i_2 i_4} \\ \delta_{i_1 i_4} \delta_{i_2 i_3} \end{pmatrix} B_{i_1}^L B_{i_2}^L = \begin{pmatrix} |\boldsymbol{B}^L|^2 \delta_{i_3 i_4} \\ B_{i_3}^L B_{i_4}^L \\ B_{i_3}^L B_{i_4}^L \end{pmatrix} \tag{46}$$

For $\boldsymbol{B}^L = \boldsymbol{y}^L$ we have $B_i^L = \delta_{iy}$ and therefore $B_{i_3}^L B_{i_4}^L = \delta_{i_3 y} \delta_{i_4 y} = \delta_{i_3 i_4} \delta_{i_3 y}$, which yields $F_{i_3 i_4}^{(4)} \propto \delta_{i_3 i_4}$ and $I_{i_3 i_4} \propto \delta_{i_3 i_4}$. The substitutions $\{a, Q, B, i_3, i_4\} \to \{d_{2,0}, \langle Q \rangle_{2,2}, \boldsymbol{y}, p, q\}$ then yield $I_{p,q}^{(2\omega)} \propto \delta_{p,q}$.

$I_{p,q}^{(\omega)}$

Assuming a linearly polarized $\boldsymbol{E}_\omega$ we must deal with an integral of the form

$$I_{i_5,i_6} = \int d\varrho [a^L \cdot B^L][b^L \cdot B^L][a^L \cdot B^L][b^L \cdot B^L]Q^L_{i_5 i_6}$$
$$= I^{(6)}_{i_1 i_2 i_3 i_4 i_5 i_6; \lambda_1 \lambda_2 \lambda_3 \lambda_4 \lambda_5 \lambda_6} a^M_{\lambda_1} b^M_{\lambda_2} a^M_{\lambda_3} b^M_{\lambda_4} Q^M_{\lambda_5, \lambda_6} B^L_{i_1} B^L_{i_2} B^L_{i_3} B^L_{i_4}, \qquad (47)$$

where

$$I^{(6)}_{i_1 i_2 i_3 i_4 i_5 i_6; \lambda_1 \lambda_2 \lambda_3 \lambda_4 \lambda_5 \lambda_6} = \int d\varrho\, l_{i_1 \lambda_1} l_{i_2 \lambda_2} l_{i_3 \lambda_3} l_{i_4 \lambda_4} l_{i_5 \lambda_5} l_{i_6 \lambda_6}, \qquad (48)$$

and we use the same notation as in the previous subsections. Using Table II in [38] we have that

$$I_{i_5,i_6} = F^{(6)}_{i_5 i_6} \cdot M^{(6)} G^{(6)}_{i_3 i_4}, \qquad (49)$$

where $\left(F^{(6)}_{i_5 i_6}\right)_r \equiv f^{(6)}_r B^L_{i_1} B^L_{i_2} B^L_{i_3} B^L_{i_4}$ and $f^{(6)}_r$ $(r = 1, 2, \ldots, 15)$ is given in Table II in [38]. For $B^L = x^L$ we have $B^L_i = \delta_{ix}$ and therefore

$$\left(F^{(6)}_{i_5 i_6}\right)_r = \begin{cases} \delta_{i_5 i_6}, & r = 1, 4, 7 \\ \delta_{i_5 x}\delta_{i_6 x}, & \text{otherwise} \end{cases} \qquad (50)$$

Since $\delta_{i_5 x}\delta_{i_6 x} = \delta_{i_5 i_6}\delta_{i_5 x}$, then $F^{(6)}_{i_5 i_6} \propto \delta_{i_5 i_6}$ and $I_{i_5,i_6} \propto \delta_{i_5 i_6}$. The substitutions $\{a, b, Q, B, i_5, i_6\} \rightarrow \{d_{1,0}, d_{2,1}, \langle Q \rangle_{2,2}, x, p, q\}$ then yield $I^{(\omega)}_{p,q} \propto \delta_{p,q}$.

References

1. E.U. Condon, Rev. Mod. Phys. **9**(4), 432 (1937). https://doi.org/10.1103/RevModPhys.9.432. URL https://link.aps.org/doi/10.1103/RevModPhys.9.432
2. K. Soai, T. Shibata, H. Morioka, K. Choji, Nature **378**(6559), 767 (1995). https://doi.org/10.1038/378767a0. URL https://www.nature.com/articles/378767a0. Number: 6559 Publisher: Nature Publishing Group
3. P. Fischer, F. Hache, Chirality **17**(8), 421 (2005). https://doi.org/10.1002/chir.20179. URL https://onlinelibrary.wiley.com/doi/abs/10.1002/chir.20179
4. S. Beaulieu, A. Ferré, R. Géneaux, R. Canonge, D. Descamps, B. Fabre, N. Fedorov, F. Légaré, S. Petit, T. Ruchon, V. Blanchet, Y. Mairesse, B. Pons, New J. Phys. **18**(10), 102002 (2016). https://doi.org/10.1088/1367-2630/18/10/102002. URL http://stacks.iop.org/1367-2630/18/i=10/a=102002
5. K. Banerjee-Ghosh, O.B. Dor, F. Tassinari, E. Capua, S. Yochelis, A. Capua, S.H. Yang, S.S.P. Parkin, S. Sarkar, L. Kronik, L.T. Baczewski, R. Naaman, Y. Paltiel, Science **360**(6395), 1331 (2018). https://doi.org/10.1126/science.aar4265. URL https://science.sciencemag.org/content/360/6395/1331. Publisher: American Association for the Advancement of Science Section: Report
6. D.S. Sanchez, I. Belopolski, T.A. Cochran, X. Xu, J.X. Yin, G. Chang, W. Xie, K. Manna, V. Süß, C.Y. Huang, N. Alidoust, D. Multer, S.S. Zhang, N. Shumiya, X. Wang, G.Q. Wang, T.R. Chang, C. Felser, S.Y. Xu, S. Jia, H. Lin, M.Z. Hasan, Nature **567**(7749), 500 (2019). https://doi.org/10.1038/s41586-019-1037-2. URL https://www.nature.com/articles/s41586-019-1037-2
7. S. Mason, Chemical Society Reviews **17**, 347 (1988). https://doi.org/10.1039/CS9881700347. URL https://pubs.rsc.org/en/content/articlelanding/1988/cs/cs9881700347
8. D.G. Blackmond, Proc. Nat. Acad. Sci. **101**(16), 5732 (2004). https://doi.org/10.1073/pnas.0308363101. URL https://www.pnas.org/content/101/16/5732

9. G.Q. Lin, Q.D. You, J.F. Cheng, *Chiral Drugs: Chemistry and Biological Action* (Wiley, Hoboken, NJ, 2011)
10. N. Berova, P.L. Polavarapu, K. Nakanishi, R.W. Woody, *Comprehensive Chiroptical Spectroscopy*, vol. 1 (Wiley, Hoboken, NJ, 2012)
11. B. Ritchie, Phys. Rev. A **13**(4), 1411 (1976). URL http://journals.aps.org/pra/abstract/10.1103/PhysRevA.13.1411
12. N. Böwering, T. Lischke, B. Schmidtke, N. Müller, T. Khalil, U. Heinzmann, Phys. Rev. Lett. **86**(7), 1187 (2001). https://doi.org/10.1103/PhysRevLett.86.1187. URL http://link.aps.org/doi/10.1103/PhysRevLett.86.1187
13. I. Powis, in *Advances in Chemical Physics* (Wiley, New York, 2008), pp. 267–329. URL http://onlinelibrary.wiley.com/doi/10.1002/9780470259474.ch5/summary
14. A.F. Ordonez, O. Smirnova, Phys. Rev. A **98**(6), 063428 (2018). https://doi.org/10.1103/PhysRevA.98.063428. URL https://link.aps.org/doi/10.1103/PhysRevA.98.063428
15. A.F. Ordonez, O. Smirnova, Phys. Rev. A **99**(4), 043416 (2019). https://doi.org/10.1103/PhysRevA.99.043416. URL https://link.aps.org/doi/10.1103/PhysRevA.99.043416
16. A.F. Ordonez, O. Smirnova, Phys. Rev. A **99**(4), 043417 (2019). https://doi.org/10.1103/PhysRevA.99.043417. URL https://link.aps.org/doi/10.1103/PhysRevA.99.043417
17. L.D. Barron, J. Am. Chem. Soc. **101**(1), 269 (1979). https://doi.org/10.1021/ja00495a071. URL http://dx.doi.org/10.1021/ja00495a071
18. J.A. Giordmaine, Phys. Rev. **138**(6A), A1599 (1965). https://doi.org/10.1103/PhysRev.138.A1599. URL https://link.aps.org/doi/10.1103/PhysRev.138.A1599
19. P.M. Rentzepis, J.A. Giordmaine, K.W. Wecht, Phys. Rev. Lett. **16**(18), 792 (1966). https://doi.org/10.1103/PhysRevLett.16.792. URL https://link.aps.org/doi/10.1103/PhysRevLett.16.792
20. D. Patterson, M. Schnell, J.M. Doyle, Nature **497**(7450), 475 (2013). https://doi.org/10.1038/nature12150. URL http://www.nature.com/nature/journal/v497/n7450/full/nature12150.html
21. D. Patterson, J.M. Doyle, Phys. Rev. Lett. **111**(2), 023008 (2013). https://doi.org/10.1103/PhysRevLett.111.023008. URL http://link.aps.org/doi/10.1103/PhysRevLett.111.023008
22. A. Yachmenev, S.N. Yurchenko, Phys. Rev. Lett. **117**(3), 033001 (2016). https://doi.org/10.1103/PhysRevLett.117.033001. URL https://link.aps.org/doi/10.1103/PhysRevLett.117.033001
23. S. Beaulieu, A. Comby, D. Descamps, B. Fabre, G.A. Garcia, R. Géneaux, A.G. Harvey, F. Légaré, Z. Mašín, L. Nahon, A.F. Ordonez, S. Petit, B. Pons, Y. Mairesse, O. Smirnova, V. Blanchet, Nat. Phys. **14**(5), 484 (2018). https://doi.org/10.1038/s41567-017-0038-z. URL https://www.nature.com/articles/s41567-017-0038-z
24. N.I. Koroteev, J. Experim. Theor. Phys. **79**(5), 681 (1994). URL http://jetp.ac.ru/cgi-bin/e/index/e/79/5/p681?a=list
25. S. Woźniak, G. Wagnière, Opt. Commun. **114**(1), 131 (1995). https://doi.org/10.1016/0030-4018(94)00498-J. URL http://www.sciencedirect.com/science/article/pii/003040189400498J
26. R. Zawodny, S. Woźniak, G. Wagnière, Opt. Commun. **130**(1), 163 (1996). https://doi.org/10.1016/0030-4018(96)00224-6. URL http://www.sciencedirect.com/science/article/pii/0030401896002246
27. B.S. Wozniak, Mol. Phys. **90**(6), 917 (1997). https://doi.org/10.1080/002689797171913. URL https://www.tandfonline.com/doi/abs/10.1080/002689797171913. Publisher: Taylor & Francis
28. P. Fischer, A.C. Albrecht, Laser Phys. **12**(8), 1177 (2002)
29. E. Gershnabel, I.S. Averbukh, Phys. Rev. Lett. **120**(8), 083204 (2018). https://doi.org/10.1103/PhysRevLett.120.083204. URL https://link.aps.org/doi/10.1103/PhysRevLett.120.083204
30. I. Tutunnikov, E. Gershnabel, S. Gold, I.S. Averbukh, J. Phys. Chem. Lett. **9**(5), 1105 (2018). https://doi.org/10.1021/acs.jpclett.7b03416. URL https://doi.org/10.1021/acs.jpclett.7b03416
31. I. Tutunnikov, J. Floß, E. Gershnabel, P. Brumer, I.S. Averbukh, Phys. Rev. A **100**(4), 043406 (2019). https://doi.org/10.1103/PhysRevA.100.043406. URL https://link.aps.org/doi/10.1103/PhysRevA.100.043406
32. A.A. Milner, J.A.M. Fordyce, I. MacPhail-Bartley, W. Wasserman, V. Milner, I. Tutunnikov, I.S. Averbukh, Phys. Rev. Lett. **122**(22), 223201 (2019). https://doi.org/10.1103/PhysRevLett.122.223201. URL https://link.aps.org/doi/10.1103/PhysRevLett.122.223201

33. P.V. Demekhin, A.N. Artemyev, A. Kastner, T. Baumert, Phys. Rev. Lett. **121**(25), 253201 (2018). https://doi.org/10.1103/PhysRevLett.121.253201. URL https://link.aps.org/doi/10.1103/PhysRevLett.121.253201

34. P.V. Demekhin, Phys. Rev. A **99**(6), 063406 (2019). https://doi.org/10.1103/PhysRevA.99.063406. URL https://link.aps.org/doi/10.1103/PhysRevA.99.063406

35. S. Rozen, A. Comby, E. Bloch, S. Beauvarlet, D. Descamps, B. Fabre, S. Petit, V. Blanchet, B. Pons, N. Dudovich, Y. Mairesse, Phys. Rev. X **9**(3), 031004 (2019). https://doi.org/10.1103/PhysRevX.9.031004. URL https://link.aps.org/doi/10.1103/PhysRevX.9.031004

36. A.F. Ordonez, O. Smirnova, [physics] (2020). URL http://arxiv.org/abs/2009.03660

37. A.F. Ordonez, O. Smirnova, [physics] (2020). URL http://arxiv.org/abs/2009.03655

38. D.L. Andrews, T. Thirunamachandran, J. Chem. Phys. **67**(11), 5026 (1977). https://doi.org/10.1063/1.434725. URL http://aip.scitation.org/doi/citedby/10.1063/1.434725

39. D.J. Cook, R.M. Hochstrasser, Opt. Lett. **25**(16), 1210 (2000). https://doi.org/10.1364/OL.25.001210. URL https://www.osapublishing.org/ol/abstract.cfm?uri=ol-25-16-1210

40. R.R. Freeman, P.H. Bucksbaum, H. Milchberg, S. Darack, D. Schumacher, M.E. Geusic, Phys. Rev. Lett. **59**(10), 1092 (1987). https://doi.org/10.1103/PhysRevLett.59.1092. URL https://link.aps.org/doi/10.1103/PhysRevLett.59.1092

2

Manipulation and Control of Molecular Beams: The Development of the Stark-Decelerator

Gerard Meijer

Abstract State-selective manipulation of beams of atoms and molecules with electric and magnetic fields has been crucial for the success of the field of molecular beams. Originally, this manipulation only involved the *transverse* motion. In this Chapter, the development of the Stark-decelerator, that allows to also manipulate and control the *longitudinal* motion of molecules in a beam, is presented.

1 Introduction

"Born in leaks, the original sin of vacuum technology, molecular beams are collimated wisps of molecules traversing the chambered void that is their theatre [...]. On stage for only milliseconds between their entrances and exits, they have captivated an ever growing audience by the variety and range of their repertoire". This is how John B. Fenn affectionately phrased it over 30 years ago, when he reflected on the long and rich history of molecular beams in his foreword to one of the classic books on this subject [1]. He could not have foreseen the spectacular leap forward that the level of control over molecular beams would take. In particular, methods that have been developed since then to slow down and store molecular beams – thereby stretching the duration of their performance on stage by orders of magnitude – have made whole new classes of experiments possible.

The motion of neutral molecules in a beam can be manipulated and controlled with inhomogeneous electric and magnetic fields. Static fields can be used to deflect or focus molecules, whereas time-varying fields can be used to decelerate or accelerate beams of molecules to any desired velocity. In this paper we present an historical overview, emphasizing the important role of molecular beam deflection and focusing experiments in the development and testing of quantum mechanics. We describe the original attempts and the successful implementation of schemes to decelerate and accelerate molecular beams with electric fields, that is, the development of the Stark-decelerator. The various elements, using electric as well as magnetic fields, that have

G. Meijer (✉)
Fritz-Haber-Institut der Max-Planck-Gesellschaft, Faradayweg 4-6, 14195 Berlin, Germany
e-mail: meijer@fhi-berlin.mpg.de

been developed for the manipulation and control of molecular beams since the first successful demonstration of the Stark-decelerator in 1998, have resulted in setups in which the molecules can be stored in stationary traps or injected in a molecular storage ring or synchrotron, for instance. Novel crossed-beam scattering studies at low collision energies, high-resolution spectroscopy studies on trapped or slow molecules and lifetime measurements of trapped molecules have become possible [2].

2   Deflection and Focusing of Molecular Beams

Atomic and molecular beams have played central roles in many experiments in physics and chemistry – from seminal tests of fundamental aspects of quantum mechanics to molecular reaction dynamics – and have found a wide range of applications [1]. Nowadays, sophisticated laser-based methods exist to perform sensitive and quantum state selective detection of the atoms and molecules in the beams. In the early days, such detection methods were lacking and the particles in the beam were detected, for instance, by a "hot wire" (Langmuir-Taylor) detector, by electron-impact ionization or by deposition and *ex-situ* investigation of the particles on a substrate at the end of the beam-machine. To achieve quantum state selectivity in the overall detection process, these methods were combined with inhomogeneous electric and magnetic field sections to influence the trajectories of the particles on their way to the detector.

The first paper discussing the degree of deflection for a beam of polar molecules passing through an inhomogeneous electric field was submitted almost a century ago, at the end of July 1921, to the *Zeitschrift für Physik* [3]. The paper was written by Hartmut Kallmann and Fritz Reiche, coworkers of Fritz Haber at the Kaiser Wilhelm Institute for Physical Chemistry and Electrochemistry in Berlin, the present Fritz-Haber-Institut der Max-Planck-Gesellschaft. Kallmann and Reiche write in their paper that they performed their analysis in support of experiments that "are ongoing at the institute" to determine whether the dipole moment is a property of an individual molecule or whether this is only induced in the molecule when it is in close proximity to other molecules, an issue that was intensively debated at that time. By passing a molecular beam, that is, a dilute but highly collimated sample of molecules, through an inhomogeneous electric field, they argued, it should be possible to monitor the deflection of the molecules and to thus determine the value of their dipole moment – provided they have one. In the introduction of their paper, they discuss in general terms the forces on a moving dipolar molecule due to inhomogeneous electric fields as well as due to inhomogeneous magnetic fields. In their analysis, however, they restrict themselves to the electric field case, and assume that the dipole moment of the diatomic molecule is perpendicular to the angular momentum vector, i.e., that the component of angular momentum along the internuclear axis is zero.

Shortly after its submission, the paper by Kallmann and Reiche came to the eyes of Otto Stern, pressing him to write up the theory behind an experiment that – as he wrote – "he had been involved in for some time with his colleague Walther Gerlach". Stern submitted his paper only about one month later, at the end of August 1921, to the same journal. In a footnote to his paper, he explicitly acknowledges that the reason for publishing his paper is the upcoming paper of Kallmann and Reiche, and concludes that both papers are nicely complementary as his paper "discusses the case in which the dipole moment is parallel to the angular momentum vector, as is generally the case for magnetic atoms" [4]. In his paper, Stern describes a method to experimentally test space quantization via measuring the deflection of a beam of atoms with a magnetic moment when moving through an inhomogeneous magnetic field [4]. No further account on the early electric deflection experiment in Berlin is to be found in the literature or in the archives, while Stern and Gerlach performed their famous experiment within one year [5], in February of 1922. Electric deflection of a beam of polar molecules was first demonstrated by Erwin Wrede, a graduate student of Stern, several years later in Hamburg [6]. It is interesting to note that up to our re-discovery of the article by Kallmann and Reiche in 2009, this article went largely unnoticed, being cited only seven times; the important mentioning of the article by Otto Stern in a footnote in his paper could not include the final reference yet. Since then, it has been cited more than once per year.

All the original experimental geometries were devised to create strong magnetic or electric field gradients to efficiently deflect particles from the beam axis. In 1939, Isidor Rabi introduced the molecular beam magnetic resonance method by using two magnets in succession to produce inhomogeneous magnetic fields of oppositely directed gradients. In this set-up the deflection of particles caused by the first magnet is compensated by the second magnet, such that the particles are directed on a sigmoid path to the detector. A transition to "other states of space quantization" induced in between the magnet sections can be detected via the resulting reduction of the detector signal. This provided a new method to accurately measure nuclear or other magnetic moments [7]. Later, both magnetic [8, 9] and electric [10] field geometries were designed to focus particles in selected quantum states onto the detector. An electrostatic quadrupole focuser, i.e., an arrangement of four cylindrical electrodes with alternating positive and negative voltages, was used to couple a beam of ammonia molecules into a microwave cavity. Such an electrostatic quadrupole lens focuses ammonia molecules that are in the so-called low-field seeking, upper level of the inversion doublet while it simultaneously defocuses those that are in the lower, high-field seeking, level. The inverted population distribution of the ammonia molecules that is thus produced in the microwave cavity led to the invention of the maser by Gordon, Zeiger and Townes in 1954–1955 [11, 12]. Apart from the spectacular observation of the amplification of microwaves by stimulated emission, these focusing elements more generally enabled the recording, with high resolution and good sensitivity, of microwave spectra in a molecular beam. By using several multipole focusers in succession, with interaction regions with electromagnetic radiation in between, versatile set-ups to unravel the quantum structure of atoms and molecules were developed. In scattering experiments, multipole focusers were exploited to study steric effects, that is, to study how the orientation of an attacking molecule affects its reactivity [13]. Variants of the molecular beam resonance methods as well as scattering machines that employed state-selectors were implemented in many laboratories, and have yielded a wealth of detailed information on stable molecules, radicals and molecular complexes, thereby contributing enormously to our present understanding of intra- and inter-molecular forces.

3   Early Attempts to Decelerate or Accelerate Molecular Beams

The state-selective manipulation of beams of atoms and molecules with electric and magnetic fields is thus about as old as the field of atomic and molecular beams itself, and it actually has been crucial for the success of the latter field. In his autobiography, Norman Ramsey, who himself later invented the separated oscillatory fields method and wrote a very influential book on molecular beam methods [14], recalls that Rabi was rather discouraged about the future of molecular beam research when he arrived in Rabi's lab in 1937, and that this discouragement only vanished when Rabi invented the molecular beam magnetic resonance method [15]. However, even though the manipulation of beams of molecules with external fields has been used extensively and with great success in the past, this manipulation exclusively involved the *transverse* motion of the molecules.

When the velocity distribution in a molecular beam is rather broad, the state-selective deflection fields can be used to provide some velocity selection. Offset or angled molecular beam geometries have been used, in which deflection fields cause only the slow (fast) atoms or molecules to obtain sufficiently large (small) deflections to pass through the apparatus and to reach the detector, for instance. This approach has been attempted to selectively load slow ammonia molecules in a microwave cavity to produce a maser with an ultra-narrow linewidth [16]. At Bell telephone laboratories an electrostatic parabolic reflector was designed to selectively couple slow ammonia molecules in the microwave cavity, that is, after deflecting them by 180 degrees [17]. These approaches suffered from the deficiency that, as stated then, "it is generally known that the velocity distribution of molecules emanating from a hole in a box is not Maxwellian, but departs from it by having fewer low velocity molecules", possibly caused by collisions with fast molecules "from behind" [17].

At the end of the nineteen-fifties, alternative approaches, with electric fields designed such as to create "multiple retardation barriers" have been proposed to actively manipulate the *longitudinal* motion of molecules in a beam, that is, to slow down ammonia molecules [18]. A rather compact experimental setup consisting out of a source chamber, a deceleration chamber and a slow-molecule deflection and detection chamber was constructed for exactly this purpose in the physics department of MIT, under the supervision of John G. King [19]. The approximately 20 cm long decelerator consisted of a linear array of ten parallel plate capacitors, capable of maintaining a voltage difference of 30 kV across 1 mm plate separation. Ammonia molecules in the low-field seeking, upper level of the $J = K = 1$ inversion doublet lose kinetic energy when entering the high electric field region of the capacitor. When the electric field is slowly turned to zero while the molecules are in the homogeneous electric field inside the capacitor, the molecules do not regain the lost kinetic energy when exiting the capacitor and the process can be repeated. The experiments were performed using continuous beams and the same high voltage was applied to the whole array of electrodes in the form of a sine wave with a fixed frequency of 6 kHz. Therefore, the distances between the adjacent parallel plate capacitors as well as their lengths needed to be made such that it takes the ever slower molecules always exactly the same amount of time to reach the next stage, i.e., these distances and lengths needed to gradually decrease along the molecular beam. The machine was designed to slow down ammonia molecules from an initial speed of 200 m/s to a final speed of 35 m/s, selectively detecting the slowed molecules at the exit. The project suffered from the same deficiency as mentioned earlier, namely that there were not enough ammonia molecules at the initial speed of 200 m/s to yield a detectable signal. The work is described in detail in the Ph.D. thesis of Robert Golub [20], but no further publication has resulted from this work.

In the physical chemistry community the experimental efforts of Lennard Wharton, to demonstrate electric field *acceleration* of a molecular beam, are much better known. In the mid nineteen-sixties, at the University of Chicago, he constructed a molecular beam machine, containing a thirty-three foot long accelerator. The accelerator consisted of an array of 600 acceleration stages intended to increase the kinetic energy of LiF molecules in high-field seeking states from 0.2 eV to 2.0 eV, that is, speeding the LiF molecules up from 1200 m/s to 3800 m/s, with the aim to use these high energy beams for reactive scattering studies. Each acceleration stage consisted of two hemispherically ended rods with a diameter of 0.5 mm, spaced 0.5 mm apart. The beam was transversely kept together using additional alternate-gradient lenses. A popular scientific account of this work, together with a schematic drawing of the acceleration principle and a photograph of the "Chemical Accelerator", appeared in *Scientific American* in 1968 [21]. The photograph of the about eleven meter long machine is reproduced in Fig. 1. Also in this case, continuous molecular beams were used and the high voltage was applied as a sine wave with a fixed frequency to the array of electrodes, implying that the adjacent electric field stages needed to be put ever further apart to compensate for the molecules being ever faster, explaining the length of the molecular beam machine. An excellent paper, in which the focusing of beams of polar molecules in high-field seeking states was theoretically analyzed, resulted from this work [22]. The acceleration experiment was not successful, however, not only because the alignment of the array of electrodes is very critical in an alternate-gradient setup but also due to an overly optimistic view on the magnitude of the electric fields that could be stably obtained when designing the accelerator module. This is what the Ph.D. student who was working on this project, Edward A. Bromberg, concluded in his Ph.D. thesis in which he summarised that "it has not been possible to show either that particles have been accelerated, or that neutral particles cannot be accelerated" [23]. Both, the deceleration experiments of John King and the acceleration experiments of Lennard Wharton were not continued after the Ph.D. students completed their theses. Whereas interest in slow molecules as a maser

Fig. 1 Photograph of the "Chemical Accelerator" built at the University of Chicago by Lennard Wharton, to produce molecular beams with an energy of about two electron volts. The photographs show the long dipole accelerator from both ends. Molecules produced by heating in an oven (left) are accelerated down the long tube by electric fields to the reaction chamber (right). Reproduced from [21]

medium declined owing to the invention of the laser, the molecular beam accelerator was made obsolete by the seminal demonstration of John Fenn and co-workers of gas dynamic acceleration of heavy species in seeded supersonic He and H_2 beams [24]. Because of these unsuccessful early attempts, researchers who wanted to pursue molecular beam deceleration with electric or magnetic fields for high-resolution spectroscopy and metrology in the following decades, had difficulties getting this financed [25].

4   Deceleration of CO ($a^3\Pi$) Molecules with Electric Fields

I studied physics at the Katholieke Universiteit Nijmegen in Nijmegen, The Netherlands – since 2004 renamed into Radboud University – and performed my undergraduate and Ph.D. research in the Atomic and Molecular Physics department, headed by Antoni Dymanus and Jörg Reuss. In this department, I was exposed from day one, that is, from February 1984 on, to a wide variety of molecular beam machines; electrostatic and magnetic state-selectors and focusing elements were used throughout. The magnetic properties and the molecular quadrupole tensor of the water molecule [26] as well as the electric and magnetic properties of carbon monoxide [27], for instance, were measured with those machines by Antoni Dymanus and his students already in the nineteen-seventies and are still the standards in the field. During my time as Ph.D. student, Jörg Reuss wrote the chapter entitled "State-selection by non-optical methods" as contribution from Nijmegen to the earlier mentioned classic book on "Atomic and Molecular beam methods" [1].

After post-doctoral periods abroad, I was offered the opportunity to start up my own research program in Nijmegen, in what was then just renamed the Molecular and Laser Physics department, and I became the successor of Jörg Reuss per January 1995. With my Ph.D. student Rienk Jongma, we performed two-dimensional imaging of metastable CO ($a^3\Pi$) molecules, to study with modern tools the phenomenon of mass-focusing in a seeded molecular beam as well as the performance of an electrostatic hexapole focuser [28]. We showed that metastable CO molecules, prepared with a pulsed laser at a well-defined time, at a well-defined position and in a single quantum state, are ideally suited to study velocity distributions and spatial distributions in molecular beams in general, and to study electric and magnetic field manipulation of molecular beams in particular. The metastable CO molecules live several milliseconds [29], long enough for molecular beam experiments, and with about 6 eV internal energy they can be efficiently detected – temporally and spatially resolved – via the Auger electrons that are released when they impinge on a surface. Whereas the Stark shift of the rotational levels of CO in its electronic ground state is very small, the shift of the $J = 1$ rotational level in the metastable $a^3\Pi_1$ state is on the order of 1 cm^{-1} in electric fields of about 100 kV/cm. When the CO molecules are state-selectively prepared in the metastable state with a pulsed laser *inside* a strong electric field, the interaction with the electric field is suddenly "switched on". This led us to propose a scheme for confining metastable CO molecules in stable "planetary" orbits in an appropriately shaped electrostatic trap [30]. As this confinement scheme would only work for CO molecules with speeds below about 22 m/s, we discussed how such slow molecules could be obtained, and in the final section of our paper we mentioned the possibility to slow down polar molecules with electric fields [30]. The latter was critically commented upon by the (anonymous) referee of our paper, which motivated us to try to experimentally demonstrate that when CO ($a^3\Pi$) molecules are prepared in a high-field seeking state *inside* a large electric field, the molecules will indeed slow down when leaving the high-field region, loosing an amount of kinetic energy that is identical to the Stark-shift of the levels. Even when seeded in Xe, however, the kinetic energy of the CO molecules in our beam was too large, to be able to unambiguously detect the deceleration effect of a single parallel-plate electric field stage of 140 kV/cm. It was clear that we would need either a considerably higher electric field or more electric field stages. When I informed Jörg Reuss in the spring of 1997 about these experiments, he told me that he vaguely remembered an experiment along similar lines by "Wharton in Chicago", but he did not know any further details. This was the first time I learned about Lennard Wharton's earlier experiments, which was comforting as I had been wondering for quite a while already why acceleration or deceleration of molecular beams with electric fields had never been demonstrated – at least now I knew it had been tried.

In December 1997, Rick Bethlem started in my department as Ph.D. student. He was hired on a project to study the physical properties of endohedral fullerenes, which would include the synthesis of small molecules like CO encapsulated in a fullerene cage, and the subsequent spectroscopic characterization of the motion of such a "dumbbell in a box". In one of our first meetings, prior to his start as Ph.D. student, I informed him about our proposed scheme to confine slow CO molecules in stable orbits and about our attempt to demonstrate molecular beam deceleration with a single electric field stage. Rick was fascinated by this topic, that actually fitted better to his background and interest, and we decided that he would switch research projects and start on the deceleration of metastable CO molecules with electric fields. At that time, I was also actively involved with other members in my department in experiments with an infrared free electron laser [31] and over lunch we had discussions about the operation principle of a linear accelerator (LINAC) for electrons, and whether and how the equivalent device for neutral, polar molecules could be constructed. We discussed in particular whether, as proposed by Basov [18], the "multiple retardation barriers" would need to be combined with electromagnetic radiation fields to be able to repeat the deceleration process, or that one could also rapidly switch the fields on and off; we concluded that both should be possible and decided to go for the approach with the time-varying electric fields. Searching for the term "slow molecules" on the internet gave us then – quite unexpectedly – as one of the first hits the one-page conference abstract by John King from 1959 in which he briefly described the ongoing ammonia deceleration experiments in his laboratory [19] (he there mentioned a 1 m long, 25-stage array of parallel-plate capacitors, using electric fields of 100 kV/cm, i.e., longer but with lower electric fields than what Robert Golub reported upon later [20]), to which we had found no reference in the later work from Lennard Wharton. Different from the earlier attempts by John King and Lennard Wharton, we used seeded pulsed beams and we did not have to rely on time-varying high voltages at a certain fixed frequency. Instead, we could make use of commercially available high voltage switches that can be rapidly (sub-μs) switched on and off in any pre-programmed time-sequence. This made it possible to design an array of – in our first experiment – 63 equidistant electric field stages with a center-to-center distance of 5.5 mm and to have complete flexibility in the input and output velocity that we would like to use. The electric field in each stage is formed by applying a high voltage to two parallel 3-mm-diameter cylindrical rods, centered 4.6 mm apart, leaving 1.6 mm opening for the molecular beam. The two opposing rods are simultaneously switched by two independent high-voltage switches from ground potential to voltages of +10 kV and −10 kV, yielding maximum electric fields of 125 kV/cm in a geometry that also provides transverse focusing. To obtain transverse focusing in two dimensions, adjacent electrode pairs are alternately positioned horizontally and vertically. All horizontal and all vertical electrode pairs are electrically connected and alternately switched, requiring a total of four independent high-voltage switches. A photograph of the prototype, 35 cm long, so-called "Stark-decelerator" as well as a scheme of the experimental set-up is shown in Fig. 2. The two electric field configurations between which switching takes place are schematically shown on the right hand side of Fig. 2.

    In Fig. 3, the original measurements are shown that gave us the first hint of signal due to decelerated molecules, indicated by the blue dashed line as a "guide to the eye" – which is needed here. Rick Bethlem had been scheduled to give an oral presentation on the deceleration experiments at the annual meeting of the Dutch AMO community in Lunteren, The Netherlands, in November 1998, and there he presented these results, that had been obtained just a few days before that. In the weeks following this meeting, we managed to significantly improve upon these first results and we demonstrated deceleration of metastable CO ($a^3\Pi_1$, $J = 1$) molecules in low-

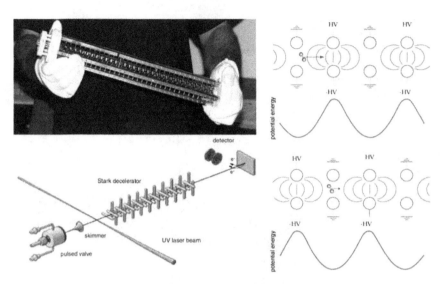

Fig. 2 Rick Bethlem, holding the prototype Stark-decelerator in his hands. The hexapole that is used to optimize the coupling of the beam into the Stark-decelerator can be seen sticking out on the left. A scheme of the experimental setup, omitting the hexapole, is shown underneath the photograph. A pulsed molecular beam is produced by expanding a mixture of 5% of CO seeded in Xe through a pulsed valve into vacuum. After passing through a skimmer, the CO molecules are excited with a pulsed laser to the low-field seeking, upper Λ-doublet component of the $J = 1$ level in the $a^3\Pi_1$ state. The metastable molecules then pass through the Stark-decelerator and are detected by monitoring the electrons that are emitted when they impinge on a clean gold surface. On the right, the two electric field configurations that are alternately used to slow down the metastable CO molecules are schematically shown.

field seeking states from 225 m/s to 98 m/s reducing their kinetic energy by almost 0.8 cm^{-1} per electric field stage [32]. Two years after this, we demonstrated phase-stable acceleration (and deceleration) of – again – CO ($a^3\Pi_1$, $J = 1$) molecules, but this time in high-field seeking states, using an array of twelve dipole lenses in alternate-gradient configuration [33]. These were the successful demonstrations of the experiments that John King and Lennard Wharton set out to perform almost forty years earlier, made possible by the choice of our system, that is, by the advantages that laser-prepared, metastable CO molecules offered, and by the advances in high-voltage switching technology.

When our first manuscript on the Stark-decelerator was still under review, I presented the main results and our future plans during a workshop at ITAMP, in July 1999. There, Daniel Kleppner informed me that although none of the attempted deceleration work of John King had been published, there should exist a Ph.D. thesis from one of his students, and we subsequently traced down the Ph.D. thesis of Robert Golub. Interestingly enough, Robert Golub mentions in his Ph.D. thesis from 1967 that "there has recently been a proposal to use a similar scheme to accelerate molecules by L. Wharton" [20]; it remains unclear in how far Lennard Wharton was

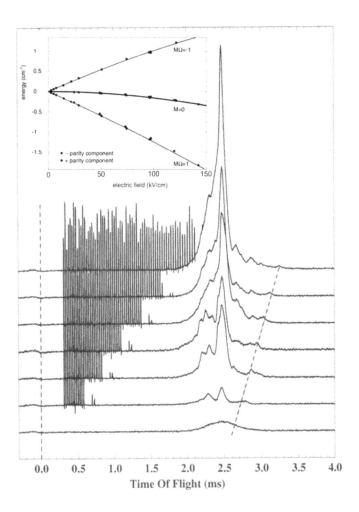

Fig. 3 Observed arrival time distribution of metastable CO molecules when an increasing number of deceleration stages is being used, from none at all (lowest curve) to all of them (uppermost curve). With no fields applied, the arrival time is centered around 2.5 ms after firing of the pulsed laser, i.e., 2.5 ms after preparation of the metastable CO molecules, with a more-or-less Gaussian distribution. The spikes in the signal prior to the arrival of the molecules on the detector are due to electrical noise from the high-voltage switches, and indicate how often the fields have been switched. When electric fields are applied, transverse focusing takes place, and more molecules are seen to arrive on the detector, with a highly structured temporal distribution. The blue, dashed line indicates the arrival time of the synchronous molecules, i.e., those CO molecules that experience the aimed-for, constant amount of deceleration per stage, and these are seen to arrive later when more deceleration stages are being used. In the inset, the measured (solid dots and squares) and calculated Stark shift of the components of the $J = 1$ level in the $a^3\Pi_1$ state of CO are shown in electric fields up to 150 kV/cm.

aware of John King's experiments. During the same workshop, Hossein Sadeghpour informed me on ongoing experiments in the group of Harvey Gould in Berkeley, aimed at decelerating molecules with electric fields. Shortly after our work was published, they published an article in which they presented data on the deceleration of Cs atoms with time-varying electric fields [34].

5 Concluding Remarks

The Stark-decelerator has made it possible to produce packets of state-selected molecules, oriented in space, with computer-controlled six-dimensional phase-space distributions. This level of control of molecular beams has first been demonstrated with electric fields, as outlined in this Chapter, but has also been obtained using time-varying magnetic fields and – to a lesser extent – using electro-magnetic radiation fields by now. Together, these methods have made a whole variety of new experiments possible. It would go too far, to (try to) list these here, and the interested reader is referred to the earlier mentioned Review from 2012 [2], and the references therein. As selected highlights that have appeared since then, I would like to mention the experimental realization of a molecular fountain by Rick Bethlem and co-workers [35], the demonstration of a cryogenic molecular centrifuge in the group of Gerhard Rempe [36], the magnetic deceleration and trapping of molecular oxygen in the group of Edvardas Narevicius [37] and the high-resolution collision experiments in the group of Bas van de Meerakker [38].

"If one extends the rules of two-dimensional focusing to three dimensions, one possesses all ingredients for particle trapping." This is how Wolfgang Paul stated it in his Nobel lecture [39], and as far as the underlying physics principles of particle traps are concerned, it is indeed as simple as that. To experimentally realize the trapping of neutral particles, however, the main challenge is to produce particles that are sufficiently slow that they can be trapped in the relatively shallow traps that can be made. When the particles are confined along a line, rather than around a point, the requirements on the kinetic energy of the particles are more relaxed, and storage of neutrons in a one meter diameter magnetic hexapole torus could thus be demonstrated first [40]. Trapping of atoms in a 3D trap only became feasible when Na atoms were laser cooled to sufficiently low temperatures that they could be confined in a quadrupole magnetic trap [41]. The Stark-decelerator enabled the first demonstration of 3D trapping of neutral ammonia molecules in a quadrupole electrostatic trap [42] even before it was used in the demonstration of an electrostatic storage ring for neutral molecules [43].

There obviously are large similarities between the manipulation of polar molecules and the manipulation of charged particles, and concepts used in the field of charged particle physics can and have been applied to neutral polar molecules, and *vice versa*. Both Hartmut Kallmann and Wolfgang Paul worked on the deflection and focusing of beams of neutral molecules before they turned their attention to controlling the motion of charged particles; it is interesting to realize that multipole fields were actually used

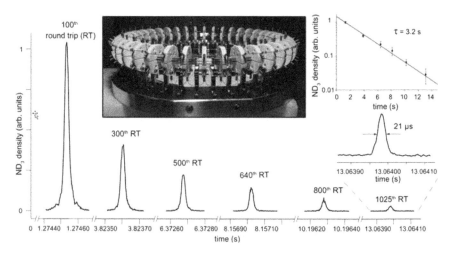

Fig. 4 Photograph of the 0.5 m diameter molecular synchrotron, consisting out of 40 straight hexapoles arranged in a circle. Deuterated ammonia molecules with a velocity of 125 m/s are injected in the synchrotron and stay confined to a 2.5 mm long packet, also after having made more than thousand round trips, i.e., after having travelled a distance of over one mile. The inset shows that the amount of trapped molecules decays with a $1/e$ time of 3.2 s, caused in equal parts by optical pumping due to black-body radiation and collisions with background gas.

in molecular beam physics first. Inspiration from charged particle physics has been instrumental for the development of the Stark-decelerator, the LINAC for neutral, polar molecules. It has also inspired the realization of a molecular synchrotron, shown in Fig. 4, in which state-selected, neutral molecules are kept together in a compact packet for a distance of over one mile, extending their duration on stage to many seconds [44].

References

1. G. Scoles, Ed., *Atomic and Molecular Beam Methods*, vol. 1 & 2 (Oxford University Press, New York, NY, USA, 1988 & 1992). ISBN 0195042808
2. S.Y.T. van de Meerakker, H.L. Bethlem, N. Vanhaecke, G. Meijer, Manipulation and control of molecular beams. Chem. Rev. **112**, 4828–4878 (2012)
3. H. Kallmann, F. Reiche, Über den Durchgang bewegter Moleküle durch inhomogene Kraftfelder. Zeitschrift für Physik **6**, 352–375 (1921)
4. O. Stern, Ein Weg zur experimentellen Prüfung der Richtungsquantelung im Magnetfeld. Zeitschrift für Physik **7**, 249–253 (1921)
5. W. Gerlach, O. Stern, Der experimentelle Nachweis der Richtungsquantelung im Magnetfeld. Zeitschrift für Physik **9**, 349–352 (1922)
6. E. Wrede, Über die Ablenkung von Molekularstrahlen elektrischer Dipolmoleküle im inhomogenen elektrischen Feld. Zeitschrift für Physik **44**, 261–268 (1927)
7. I.I. Rabi, S. Millman, P. Kusch, J.R. Zacharias, The molecular beam resonance method for measuring nuclear magnetic moments. Phys. Rev. **55**, 526–535 (1939)
8. H. Friedburg, W. Paul, Optische Abbildung mit neutralen Atomen. Die Naturwissenschaften. **38**, 159–160 (1951)
9. H.G. Bennewitz, W. Paul, Eine Methode zur Bestimmung von Kernmomenten mit fokussiertem Atomstrahl. Zeitschrift für Physik **139**, 489–497 (1954)

10. H.G. Bennewitz, W. Paul, C. Schlier, Fokussierung polarer Moleküle. Zeitschrift für Physik **141**, 6–15 (1955)

11. J.P. Gordon, H.J. Zeiger, C.H. Townes, Molecular microwave oscillator and new hyperfine structure in the microwave spectrum of NH_3. Phys. Rev. **95**, 282–284 (1954)

12. J.P. Gordon, H.J. Zeiger, C.H. Townes, The maser—new type of microwave amplifier, frequency standard, and spectrometer. Phys. Rev. **99**, 1264–1274 (1955)

13. R. Levine, R. Bernstein, *Molecular Reaction Dynamics and Chemical Reactivity* (Oxford University Press, New York, 1987)

14. N.F. Ramsey, *Molecular Beams*, in The International Series of Monographs on Physics (Oxford University Press, London, 1956)

15. N.F. Ramsey, http://nobelprize.org/nobel_prizes/physics/laureates/1989/ramsey-autobio.html

16. D.C. Lainé, Molecular beam masers. Rep. Prog. Phys. **33**, 1001–1067 (1970)

17. L.D. White, Ammonia maser work at Bell telephone laboratories, in *Proceedings of the 13th Annual Symposium on Frequency Control* (Fort Monmouth, U.S. Army Signal Research and Development Laboratory, 1959), pp. 596–602

18. N.G. Basov, A.N. Oraevskii, Use of slow molecules in a maser. Soviet Phys. JETP **37**, 761–763 (1960)

19. J.G. King, Experiments with slow molecules, in *Proceedings of the 13th Annual Symposium on Frequency Control* (Fort Monmouth, U.S. Army Signal Research and Development Laboratory, Asbury Park, 1959), p. 603

20. R. Golub, *On Decelerating Molecules* (Ph.D. thesis, MIT, Cambridge, USA, 1967)

21. R. Wolfgang, Chemical accelerators. Sci. Am. **219**(4), 44–52 (1968)

22. D. Auerbach, E.E.A. Bromberg, L. Wharton, Alternate-gradient focusing of molecular beams. J. Chem. Phys. **45**, 2160–2166 (1966)

23. E.E.A. Bromberg. *Acceleration and Alternate-Gradient Focusing of Neutral Polar Diatomic Molecules*. Ph.D. thesis, University of Chicago, USA (1972)

24. N. Abuaf, J.B.A.R.P. Andres, J.B. Fenn, D.G.H. Marsden, Molecular beams with energies above one volt. Science **155**, 997–999 (1967)

25. E.A. Hinds, Private communication (Heidelberg, 1999)

26. J. Verhoeven, A. Dymanus, Magnetic properties and molecular quadrupole tensor of the water molecule by beam-maser Zeeman spectroscopy. J. Chem. Phys. **52**, 3222–3233 (1970)

27. W.L. Meerts, A. Dymanus, Electric and magnetic properties of carbon monoxide by molecular-beam electric-resonance spectroscopy. Chem. Phys. **22**, 319–324 (1977)

28. R.T. Jongma, Th Rasing, G. Meijer, Two-dimensional imaging of metastable CO molecules. J. Chem. Phys. **102**, 1925–1933 (1995)

29. J.J. Gilijamse, S. Hoekstra, S.A. Meek, M.Metsälä, S.Y.T. van de Meerakker, G. Meijer, G.C. Groenenboom, The radiative lifetime of metastable CO ($a^3 \Pi$,v=0). J. Chem. Phys. **127**, 221102-1–221102-4 (2007)

30. R.T. Jongma, G. von Helden, G. Berden, G. Meijer, Confining CO molecules in stable orbits. Chem. Phys. Lett. **270**, 304–308 (1997)

31. G. von Helden, I. Holleman, G.M.H. Knippels, A.F.G. van der Meer, G. Meijer, Infrared resonance enhanced multiphoton ionization of fullerenes. Phys. Rev. Lett. **79**, 5234–5237 (1997)

32. H.L. Bethlem, G. Berden, G. Meijer, Decelerating neutral dipolar molecules. Phys. Rev. Lett. **83**, 1558–1561 (1999)

33. H.L. Bethlem, A.J.A. van Roij, R.T. Jongma, G. Meijer, Alternate gradient focusing and deceleration of a moleclar beam. Phys. Rev. Lett. **88**, 133003-1–133003-4 (2002)

34. J.A. Maddi, T.P. Dinneen, H. Gould, Slowing and cooling molecules and neutral atoms by time-varying electric-field gradients. Phys. Rev. A **60**, 3882–3891 (1999)

35. C. Feng, A.P.P. van der Poel, P. Jansen, M. Quintero-Pérez, T.E. Wall, W. Ubachs, H.L. Bethlem, Molecular fountain. Phys. Rev. Lett. **117**, 253201-1–253201-5 (2016)

36. X. Wu, T. Gantner, M. Koller, M. Zeppenfeld, S. Chervenkov, G. Rempe, A cryofuge for cold-collision experiments with slow polar molecules. Science **358**, 645–648 (2017)

37. Y. Segev, M. Pitzer, M. Karpov, N. Akerman, J. Narevicius, E. Narevicius, Collisions between cold molecules in a superconducting magnetic trap. Nature **572**, 189–193 (2019)

38. T. de Jongh, M. Besemer, Q. Shuai, T. Karman, A. van der Avoird, G.C. Groenenboom, S.Y.T. van de Meerakker, Imaging the onset of the resonance regime in low-energy NO-He collisions. Science **368**, 626–630 (2020)
39. W. Paul, Electromagnetic traps for charged and neutral particles. Angew. Chem. Int. Ed. Engl. **29**, 739–748 (1990)
40. K.-J. Kügler, W. Paul, U. Trinks, A magnetic storage ring for neutrons. Phys. Lett. B. **72** 422–424 (1978)
41. A.L. Migdall, J.V. Prodan, W.D. Phillips, T.H. Bergeman, H.J. Metcalf, First observation of magnetically trapped neutral atoms. Phys. Rev. Lett. **54**, 2596–2599 (1985)
42. H.L. Bethlem, G. Berden, F.M.H. Crompvoets, R.T. Jongma, A.J.A. van Roij, G. Meijer, Electrostatic trapping of ammonia molecules. Nature **406**, 491–494 (2000)
43. F.M.H. Crompvoets, H.L. Bethlem, R.T. Jongma, G. Meijer, A prototype storage ring for neutral molecules. Nature **411**, 174 (2001)
44. P.C. Zieger, S.Y.T. van de Meerakker, C.E. Heiner, H.L. Bethlem, A.J.A. van Roij, G. Meijer, Multiple packets of neutral molecules revolving for over one mile. Phys. Rev. Lett. **105**, 173001-1–173001-4 (2010)

The Precision Limits in a Single-Event Quantum Measurement of Electron Momentum and Position

H. Schmidt-Böcking, S. Eckart, H. J. Lüdde, G. Gruber and T. Jahnke

Abstract A modern state-of-the-art *"quantum measurement"* [The term "quantum measurement" as used here implies that parameters of atomic particles are measured that emerge from a single scattering process of quantum particles.] of momentum and position of a **single** electron *at a given time* ["at a given time" means directly after the scattering process. (It should be noticed that the duration of the reaction process is typically extremely short => attoseconds).] and the precision limits for their experimental determination are discussed from an **experimentalists point of view**. We show—by giving examples of actually performed experiments—that in a single reaction between quantum particles *at a given time* only the momenta of the emitted particles but not their positions can be measured with sub-atomic resolution. This fundamental disparity between the conjugate variables of momentum and position is due to the fact that during a single-event measurement only the total momentum but not position is conserved as function of time. We highlight, that (other than prevalently perceived) Heisenberg's "Uncertainty Relation" UR [1] does not limit the achievable resolution of momentum in a **single-event measurement**. Thus, Heisenberg's statement that in a single-event measurement only either the position or the momentum (velocity) of a quantum particle can be measured with high precision contradicts a real experiment. The UR states only a correlation between the mean statistical fluctuations of a large number of repeated single-event measurements of two conjugate variables. A detailed discussion of the real measurement process and its precision with respect to momentum and position is presented.

H. Schmidt-Böcking (✉) · S. Eckart · G. Gruber · T. Jahnke
Institut für Kernphysik, Universität Frankfurt, 60438 Frankfurt, Germany
e-mail: hsb@atom.uni-frankfurt.de; schmidtb@atom.uni-frankfurt.de

H. J. Lüdde
Institut für Theoretische Physik, Universität Frankfurt, 60438 Frankfurt, Germany

1   Introduction

Otto Stern was the pioneer in high-resolution momentum spectroscopy of atoms and molecules moving in vacuum. Gerlach and Stern performed between 1920 to 1922 in Frankfurt their famous Stern-Gerlach experiment (SGE). They obtained for Ag atoms a sub-atomic momentum resolution in the transverse direction of 0.1 a.u. [2]. Today, modern state-of-the-art spectrometer devices such as the Scienta electron spectrometers [3] or the COLTRIMS Reaction Microscope C-REMI [4] can provide even a much better resolution. The imaging system C-REMI can even measure several particles in coincidence by detecting the momenta of all charged fragments emitted in a quantum process. Thus the complete entangled dynamics of such a single quantum process can be visualized. However, in such high resolution experiments the experimenter cannot obtain any direct information on the relative positions of particles. For a single event the absolute and also relative positions inside the quantum reaction are not measureable. The purpose of this paper is to illustrate for a single-event scattering measurement, as discussed by Heisenberg [1], the precision limits of electron momentum and position by presenting experimental examples.

The goal of a quantum measurement, e.g. scattering of a quantum projectile on a target atom in vacuum, is to obtain information on the quantum mechanical collision process. How can such a measurement be performed? The experimenter must prepare projectiles and target objects in a well-defined momentum and as far as possible also in a well-defined position state. This is typically achieved by classical methods. As shown below the momentum state of projectile and target object can be prepared with sub-atomic precision, but positions at a given time can never be controlled with atomic size accuracy. The reason is, no particle in vacuum can be brought completely at rest in the system of measurement and thus positions are not conserved with time. In other words the experimenter cannot predict with sub-atomic precision the impact parameter of the collision and the impact parameters are statistically distributed. Thus numerous single event measurements one after the other have to be summed up to obtain a statistical distribution.

The statistical distribution contains two sources of errors: First, the systematical error of each single-event measurement. This is de facto the horizontal error bar which is given by the quality of preparation and of the classical detection device only. This error bar depends very little on Heisenberg's uncertainty relation. Second, there is a statistical error in the ordinate values, which depends on the number of detected events and which does not depend on the precision of the single-event measurement. The sum of all single-event measurements, i.e. the statistical distribution, is relevant for comparison with theory.

What are the precision limits for parameters in the quantum experiment? The detection apparatus delivers only auxiliary values, from which then information on the quantum process can be deduced. Such auxiliary values are: The time, when the collision occurs. It can be determined by classical methods (see below) with about 50 pico-second precision. During the collision electrons and ionic fragments can be emitted each with a so-called final momentum. Immediately after emission they move in a spectrometer device in which the charged quantum particles exchange momentum due to the electro-magnetic force with the macroscopic detection device. Finally the particles impact on a detector, which can be placed at any distance from the collision zone. The auxiliary values, that are measured by the detector, are: detector impact position and time. Typically they are measured with a precision of 50 μm and 50 pico-seconds. It is to be noticed that these auxiliary quantities allow the experimenter to deduce the particle trajectories in the detection device and to determine the final momentum in the laboratory system from the trajectory of the particle (see below).

To obtain sub-atomic momentum precision (laboratory system) in a single event, the velocity vector (i.e. the total momentum) of the center-of-mass of the single-event collision system must be known from the method of preparation. Conservation laws are therefore of fundamental importance for the implementation of a single-event quantum measurement. An observable can only be measured with sub-atomic accuracy if time-dependent conservation properties are strictly fulfilled during the generally very short duration of the measurement. Total linear momentum, total angular momentum and total energy are conserved but not location. The measured momenta of all fragments can then be corrected for the center-of-mass motion because the total momentum is conserved. For position no conservation law exists, thus a large uncertainty in the location measurement cannot be avoided. Therefore Heisenberg's suggestions that a high resolution position measurement is possible and this position measurement would be even the basis of any quantum measurement completely contradicts real experiments.

Since 1927 numerous papers have been published discussing the consequences of the UR on a quantum measurement within the wave-picture. To the best of our knowledge there is no publication available, where the constraints and the purely classical experimental limits of a single-event quantum measurement are analyzed from the view of an experimenter. Although in the introduction of his paper [1], Heisenberg considered the kinematics and mechanics of a single particle and the measurement of the position and the velocity (momentum) of a single electron "at a given moment", Heisenberg's UR ($\Delta x \cdot \Delta p \geq \hbar$) applies, however, only for the mean statistical fluctuations of a large number of repeated single-event measurements of two conjugate variables and can be viewed to be a prediction of the future particle properties.

We deploy therefor the following two statements:

Statement 1. The UR applies for the statistical distribution of a large ensemble or for repeated measurements but not for the resolution of a single-event measurement.

This statement is in line with previous work, that revisited this discussion, as well. For example, Ballentine [5], Park and Margenau [6] as well as Briggs [7, 8] contradicted the single-event interpretation of Heisenberg and concluded that it applies only to a large ensemble of similarly prepared systems.

Statement 2. A single-event measurement can only provide information on the particle's properties back in the past but never allow a prediction of future properties, since the impact of a particle on the detector changes the particles momentum and position state.

Which parameters of a quantum reaction are measurable and what is the achievable precision? We discuss this question by illustrating the concept of a "real" quantum experiment. As paradigm example for a typical quantum measurement we have chosen the scattering of an electron or ion on a gaseous target atom followed by the coincident detection of all reaction fragments with modern "state-of-the-art" detection devices.

We will discuss the following three findings:

1. **One can measure the final momenta of all emitted charged fragments.** Since each single-event measurement takes some time (from preparation until detection), the conservation with time of the total momentum of the whole scattering system is a crucial property in order to obtain excellent resolution in real measurements. During the short period of measurement the momenta of all particles, are "correlated" due to the law of momentum conservation, i.e. they are even dynamically entangled, for the whole time until they finally impact on a classical detector (see Ref. [7] and comments therein connected to this paper).

2. The angular momentum of a single freely-moving electron emitted in a quantum reaction appears undetectable. However, the quantum states (whose quantum numbers) can be deduced, if the electron kinetic energy can be assigned to a well-defined transition. In an ion-atom collision process, however, a coincidence measurement can provide information on the **angular momentum** of a single particle. In the case of a complete multi-particle coincidence measurement, when the nuclear collision plane is determined, this additional information can be employed in some cases to deduce the angular momentum, as, for example, certain angular momentum states are emitted due to space quantization only into distinct regions like in the Stern-Gerlach experiment (see e.g. data in Fig. 4 of this paper).

3. One can also precisely determine the **amount of the electronic excitation energy** from the measured momenta of all particles in the preparation and final states, because the total energy is also conserved (assumption: projectile and target in the preparation state are in the ground-state). The excitation energy is then the difference between the kinetic energies in the initial and final states.

The UR imposes, in contradiction to Heisenberg's claim, no limit on the achievable momentum (velocity) resolution of a single quantum measurement. The UR affects the resolution of such a measurement only indirectly, as it has an impact on the quality of preparation of the pre-collision states of projectiles and target atoms. This has already been highlighted by Kennard in 1927 [9] who theoretically considered the passage of scattered electrons in a classical detection device and concluded:

*„In den hier behandelten Fällen haben wir keinerlei quantentheoretische Abwe-
ichung gefunden von den klassischen Ergebnissen. Die einzige quantenhafte Eigen-
tümlichkeit in solchen einfachen Fällen liegt in der durch das Heisenberg'sche Unbes-
timmtheitsgesetz festgesetzten prinzipiellen Unbestimmtheit der Anfangswerte"* (*"In
the cases discussed here, we did not find any quantum theoretical deviation from the
classically calculated values. The only quantum influence in such cases originates
from the effect that the preparation state values are indeterminate in accordance to
Heisenberg's Uncertainty Relation."*). Today, the debate over the statistical versus
single-event interpretation is still not converged (see [5–16] for proponents of the
single-event interpretation and for papers opposing this interpretation).

In the following chapters we discuss the **purely experimental aspects** and
the limits of experimental precision in a single-event measurement of momentum
(velocity) and position and present examples:

In Sect. 2: The **scheme and time evolution of a single-event measurement**
is discussed beginning with the preparation of the measurement followed by
the quantum reaction process and concluding with the detection of the charged
fragment in a classical measurement apparatus.

In Sect. 3: The **electron momentum (velocity) measurement by Time-of-
Flight (TOF) trajectory imaging** is presented. We consider realistic experimental
scenarios for electrons based on experimental results.

In Sect. 4: The determination of the **angular momentum state** of a single electron
by a multi-fragment coincidence technique.

In Sect. 5: The **experimental limits for an electron position measurement**
are discussed. We also show that Heisenberg's "Gedankenexperiment" on the
γ-microscope is not feasible.

In Sect. 6: We consider the **product of precisions in momentum and position
measurement of a freely moving single electron**. New experimental techniques
for measuring momentum and position of a freely moving electron simultaneously
in a one-step approach are provided for the moment of impact on a detector. Within
this approach the product of the experimental error bars in electron momentum
and detector impact position can be below ħ by several orders of magnitude.

2   Scheme of a Quantum Measurement

We consider an experiment where a projectile beam intersects in ultra-high vacuum
with a gaseous target to ensure controlled single-event conditions i.e. that only one
reaction process occurs during each measurement period. Because of the statistical
nature of quantum measurements (to yield statistical distributions) one must prepare
numerous projectiles in the "nearly identical" pre-collision state and numerous target
objects in controlled "nearly identical" momentum and position states. In the preparation of the pre-collision state "nearly identical" means this preparation is still
limited by Heisenberg's UR with respect to the large ensemble projectile and target

momentum and position fluctuation widths. E.g. in an ion-atom collision the experimenter cannot precisely adjust the impact parameter to obtain the same deflection angle. The selection of impact parameters is of pure statistical nature. Thus, the experiment has to be repeatedly performed with numerous of such single projectiles and target objects. Finally summing over a huge number of single-events the experimenter obtains a statistical distribution that allows for the retrieval of the final-state fluctuation width (with the help of theory also quantum mechanical properties or properties of the wave function).

2.1 Time Evolution of a Quantum Measurement

In Fig. 1 the scheme of a single-event quantum experiment and the time evolution of such a complete quantum measurement process are shown. The measurement may be separated in three sequential steps: the time of preparation (pre-collision step, zone A), the time of reaction (zone B), and the time after the reaction (post-collision step, zone C) before the reaction products impact on the detector. In the view of the experimenter the momenta and trajectories of the particles in the macroscopic preparation stage A (pre-collision) as well as in the macroscopic spectrometer system C (post-collision) can be treated by the laws of classical physics. The very tiny reaction region B (typically of atomic to micrometer size) is a purely quantum mechanical region and must be treated accordingly. The dynamics in region B cannot be directly observed by the experimenter. The classical behavior in A and C is justified theoretically by the Imaging Theorem of the accompanying papers [7, 8]. This result shows

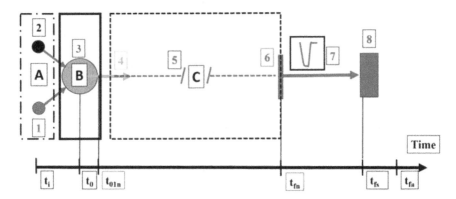

Fig. 1 Time evolution of a quantum measurement. A indicates the time interval before the interaction of projectile (1) and target (2), B is the very short time interval of the quantum scattering process (3) (occurring at the time t_0) and C the time interval in which the emitted reaction particles (4) are travelling inside the classical detection setup. The particle is finally detected on a detector (6). The detector yields an electronic signal (7) (typically a nanosecond long) providing time information on the quantum scattering event, which is stored electronically in a computer (8)

that, after propagation to or from macroscopic distances, the position and momentum variables of the quantum wave function obey classical relations.

The reaction products emitted in the quantum reaction are interacting with the macroscopic measurement apparatus in zone C. In the macroscopic apparatus they can be treated as classical particles with classically defined momenta (moving on classical trajectories) since they exchange in zone C de facto only momentum with the measurement device due to applied electric or magnetic spectrometer fields. Any interaction of the fragments with the rest gas in the spectrometer can be excluded because of the very low vacuum pressure (typically below 10^{-8} millibar). At the end of the macroscopic detection device position-sensitive detectors measure the impact position in the laboratory system of each fragment and also the time of impact for each fragment separately (if required all fragments can be measured in coincidence).

As Popper pointed out [17], after completion of a measurement the experimenter determines always the kinematical parameters of the "past" for each single event, whereas the UR makes predictions into the future for the outcome of statistical distributions of many repeated single-event measurements.

3 Electron Momentum (Velocity) Measurement by Time-of-Flight (TOF) Trajectory Imaging

3.1 The Experimental Scheme for Momentum (Velocity) Measurement

In the following we describe a quantum measurement of charged particles from an ionization process using a momentum-imaging approach. After leaving the reaction zone B (see Fig. 1 at time t_{01n}) the charged fragments begin to move in zone C on "quasi-classical" trajectories (see Refs. [7, 8]) with classically defined momenta, since in zone C they nearly exclusively exchange momentum with the spectrometer via classical forces. The distance d from the reaction point, from where one can neglect quantum mechanical post-collision interaction, can be crudely estimated by comparing the strengths of interacting forces, i.e. the magnitude of momentum exchange. In zone B the force between electron and ion dominates and in zone C the force imposed on the charged particles by the spectrometer fields is dominating. This is because the force between electron and ion depends on their distance d. Assuming the ion is singly charged then the electron-ion force is $F_{ion} = e^2/d^2$ (in a.u.). For d = 1000 a.u. one obtains $F_{ion} = 10^{-6}$ a.u., for d = 1 μm one obtains $F_{ion} = 2.8 \times 10^{-9}$ a.u. The strength of the classical force in the fields of the measurement device can be estimated from the electric field strength in the spectrometer. The field is typically larger than 10 V/cm, thus for an electron the acting spectrometer force is $F_{eS} > e \cdot 10$ V/cm $= 1.92 \times 10^{-9}$ a.u. Therefore, for distances d larger than a few tens of micrometers the electron-ion force strength can be neglected and the

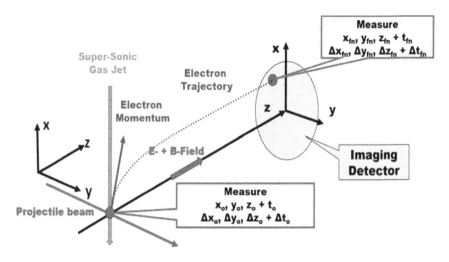

Fig. 2 Scheme of trajectory imaging technique for charged quantum particles in a classical spectrometer [4]. The electron momentum vector (blue arrow) is the so-called "final momentum", with which the electron is emitted from the collision process with respect to the center-of-mass system of the reaction process

emitted fragments are only interacting with the spectrometer field yielding a well-defined classical trajectory due to momentum conservation (charged fragment plus spectrometer are entangled). The momentum change and thus the classical trajectory of the fragment in the spectrometer depend on the electric-magnetic field design and on the final fragment momentum p_{fn}.

A static electric field accelerates electrons and positively charged ionic fragments into opposite directions. The fragments are finally detected by two position- and time-sensitive detectors placed in opposite directions (only one direction is shown in Fig. 2). Since the spectrometer provides for positively and negatively charged particles nearly a 4π-detection efficiency it can capture a complete image of the reaction process in momentum space.

The measurement of the final momentum of an emitted fragment can thus be achieved through a precise determination of the particle trajectory in part C in the classical detection device. To determine the complete classical trajectory of each particle one has to measure only the classical location parameters $r_0 = (x_0, y_0, z_0)$ and $r_{fn} = (x_{fn}, y_{fn}, z_{fn})$ as well as times t_0 and t_{fn} (see Fig. 2). Both time parameters can be determined with a precision of about 50 pico-seconds, t_0 can be measured by using a timed-bunched projectile beam and t_{fn} by using a "state-of-the-art" classical detection device [4]. Target location and position of impact on the detector can be measured with a precision of better than 50 μm (even 10 μm are achievable). Knowing (or calibrating) the electro-magnetic field configuration and measuring the above listed parameters, the final momentum vector of the fragment can easily be deduced by using simple classical equations [4]. **Although all auxiliary parameters**

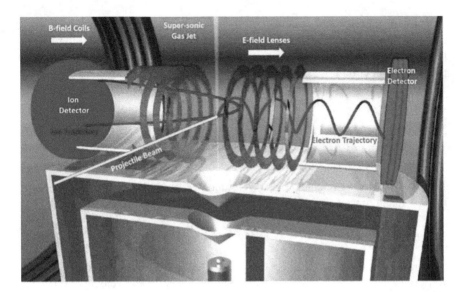

Fig. 3 Scheme of the C-REMI [4] which can image with 4π solid angle all emitted charged particles (ions: red trajectory, electrons: blue trajectory) in coincidence. A projectile beam intersects in the center of the C-REMI with a super-sonic gas jet (from below) inducing the quantum reaction process. The applied electric field super-imposed by a magnetic field (see the brown coils) projects all charged fragments/electrons on position- and time-sensitive detectors

are measured with macroscopic accuracy only, sub-atomic resolution for the electron and ion momenta (velocities) can be obtained.

The C-REMI [4] is such a "state-of-the-art" momentum-imaging device. In Fig. 3 the scheme of such a detection approach is presented. The reaction takes place within the tiny intersection region of projectile and target beams (e.g. internally very cold super-sonic gas jet). The blue and red curves in Fig. 3 indicate the classical trajectories of ionic fragments (red line) and electrons (blue line) in the spectrometer. With the help of electric and magnetic fields nearly all fragments are projected on position-sensitive detectors yielding a very high multi-coincidence detection efficiency.

Before we discuss a real experimental scenario, we first define "good" and "bad" resolution in a single-event quantum measurement with respect to the standard dimensions in an atomic system. The standard sizes of atomic parameters are defined by the classical features of an electron in a hydrogen atom. The classical K-shell radius is $r_K = 5.29 \times 10^{-9}$ cm, which is used to define the atomic unit of length (a.u.). The classical electron velocity of the electron in the hydrogen K-shell is $v_K = 2.18 \times 10^8$ cm/s, which defines 1 a.u. of velocity. The classical momentum of the electron in the hydrogen K-shell is $p = m_e v_K = 1$ a.u. An atomic unit of time is defined by the ratio of the hydrogen K-shell radius divided by the corresponding electron velocity, or 5.29×10^{-9} cm divided by 2.18×10^8 cm/s yielding 24 attoseconds. Furthermore, the electron charge e and mass are also set to 1 a.u. and hence $\hbar$ results to be 1 a.u., too.

Thus, it appears very reasonable when we define resolution of single-event quantum measurements with respect to these atomic units. "Good" sub-atomic resolution is on the order of a few percent of one a.u. and "very good" resolution is on the order of a per mill or even better. Bad resolution is larger than one a.u.

3.2   Momentum (Velocity) Measurement and Its Achievable Resolution for an Electron

The achievable experimental precisions for momentum (velocity) are discussed here for two quantum processes. First, the transfer ionization process which is

$$10 \, keV \, He^{2+} + He => He^{1+} + He^{2+} + e$$

investigated by Schmidt et al. [18]. This experiment was performed to search for vortices in the electron current which should be visible in the velocity/momentum distribution of the emitted electrons. To visualize such effects in the electron momentum distribution, a high experimental momentum resolution ($\delta p = 0.01$ a.u.) in a single event is required. Additionally, a coincidence measurement with the ejected ions is necessary in order to determine the orientation of the quasi-molecule during the collision. This was achieved with the C-REMI approach. During such slow collisions quasi-molecular orbitals are formed and electrons are promoted to the continuum via a few selected angular momentum states.

In Fig. 4 the measured electron-momentum distributions are shown together with the achieved single-event resolution δp (black square) and with one example of a momentum fluctuation width $<\Delta p>$ (varies with electron energy). It is to be noticed that in this experiment of Schmidt et al. the electron-detector distance from the intersection region (gas jet-projectile beam) was only 3 cm due to other experimental requirements. This short distance limits the momentum resolution, because of the very short TOF. Nevertheless a resolution of 0.01 a.u. was obtained. The resolution can be improved by increasing this distance.

To demonstrate the high resolving power for electron momenta of the C-REMI a numerical example, a kind of "*Gedankenexperiment*", i.e. the process of electron impact ionization of He

$$e + He => He^{1+} + 2e$$

is discussed here. Today such an experiment would be feasible.

In Appendix A the preparation of the required electron beam quality is described which enables the high required momentum accuracy of the projectiles. To yield the required excellent "Time-of-Flight" TOF resolution the detectors should be located as far as possible from the zone B, i.e. the spectrometer should be as large as possible. For a trajectory length inside region C (from zone B to the electron detector surface)

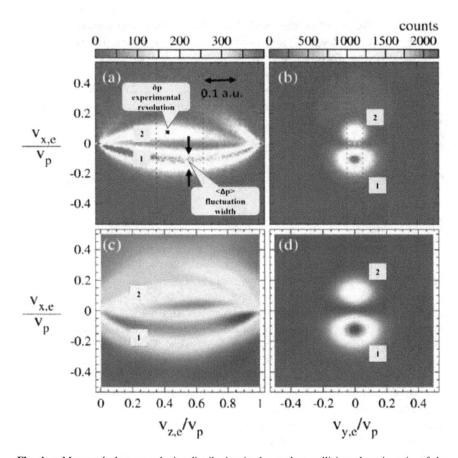

Fig. 4 **a** Measured electron velocity distribution in the nuclear collision plane in units of the projectile velocity $v_p = 0.63$ a.u. for small nuclear scattering angles <1.25 mrad. **b** Perpendicular to the nuclear collision plane. **c, d** corresponding theoretical predictions. An electron moving with the projectile velocity $v_p = 0.63$ a.u. has a momentum of 0.63 a.u. [18]. The experimental resolution in a single event of $\delta p = 0.01$ a.u. corresponds to an energy resolution of approximately 1 meV

of 2 m the angular resolution of the trajectory measurement is of the order of the sum of the intersection width of projectile beam and target beam and detector position resolution divided by the trajectory length. This ratio is about $2 \cdot 50$ μm/200 cm $\approx 0.5 \times 10^{-4}$. This geometrical ratio limits the transverse momentum resolution in x- and y-direction. The longitudinal momentum resolution (in z-direction) depends on the TOF resolution. An electron moving with 2 a.u. momentum has a velocity of $4.38 \times 10^{+8}$ cm/sec and its total TOF inside C is 200 cm/$4.38 \times 10^{+8}$ (cm/sec) $= 450$ ns. Thus the relative TOF resolution $\Delta TOF/TOF$ is about 10^{-4} yielding an overall momentum precision for an electron of 2 a.u. momentum of 2×10^{-4} a.u..

We would like to notice, that in C-REMI the velocities and masses of moving particles are measured, which yield directly the momenta. The velocities are macro-scopically large and therefore directly measurable with macroscopic classical TOF devices. Heisenberg considered the measurement of the velocity (momentum) of an electron bound in an atom too. His approach will be discussed in Appendix B together with the possibility of momentum measurements of bound electrons via the process of Compton scattering.

4   Measurement of Angular Momentum of a Single Electron

Any bound electron usually has an orbital angular momentum in addition to its own spin. Due to the spin-orbit coupling, all electrons in an atom form one unit providing a quantized total angular momentum. If an experimenter can only measure the momentum of only one emitted electron (so-called single parameter measure-ment), then an experimenter can hardly make any statement about the quantum state in which the electron was originally bound. In case of single parameter measurement only from the electron momentum distribution of a large amount of identical ioniza-tion processes one can make a statement about the type of multipole distribution and thus on the angular momentum transfer involved. Thus the angular momentum of a single freely-moving electron emitted in a quantum reaction appears undetectable.

However, if the electron kinetic energy can be assigned to a single transition between well-defined quantum states, whose quantum numbers can be deduced. Furthermore in an ion-atom collision process and in the case of a complete multi-particle coincidence measurement, when the nuclear collision plane is determined too, this additional information can be employed in some cases to deduce the angular momentum states of a single ejected electron. In a slow ion-atom collision process, quasi-molecular electronic orbitals are formed during the collision, which are sharply angularly quantized with respect to the nucleus-nucleus scattering plane. Thus different angular momentum states are emitted due to space quantization only into distinct regions like in the Stern-Gerlach experiment (see e.g. data in Fig. 4 of this paper). If e.g. in a transfer-ionization process an electron passes over from these quasi-molecular states into the continuum [18] then the electrons in the x-y plane perpendicular to the nucleus-nucleus scattering plane are emitted with discrete transverse momenta (Fig. 4) and the different quasi-molecular orbitals e.g. 1 and 2 in Fig. 4 can clearly distinguished. Just as in the Stern-Gerlach experiment, these discrete transverse momenta correspond to certain angular momentum states which can be discerned in a coincidence measurement.

This clearly proves (Fig. 4: comparison of experiment and theory) that in a coinci-dence experiment the directional quantization of the quasi-molecular states becomes measureable and thus in selected collision systems the angular momentum states of single emitted electrons can be determined too.

5   Electron-Position Measurement and Achievable Resolution

Heisenberg described the position measurement of single electrons at a given moment as the foundation of any parameter measurement. He proposed to measure the velocity by detecting the electron positions at two succeeding moments. He explained his view on position measurements by thought experiments: *"If one wants to under-stand, what the definition of 'position of a particle', e.g. of the electron (relative to the reference system of measurement) means, one must describe well-defined experimental approaches, how the 'position of an electron' can be measured; otherwise the definition of position is meaningless.* He continued: ***"There is no shortage of such experimental approaches, which can measure the 'position of an electron' with unlimited precision."*** (page 174) [1]. Therefore he viewed a trajectory as a discontinuous path because of discontinuous observations. On page 185 he continued: *"I believe that the appearance of a classical trajectory is manifested by its observation"*.

Heisenberg proposed to use a so-called γ-microscope to measure the position of a quantum object, e.g. an electron at a given moment. He ascertained [1]: *"The resolution of the light microscope is only limited by the wave length of the light. Using short wave length x-rays the resolution should have no limitation."* The scheme of such a photon microscope measurement can only be explained in the wave-picture (thus many photons must be detected). But one has to make sure that the object is not changing its position during the exposure time of the measurement. With the help of such a microscope (combination of lenses) one can magnify tiny objects and project their image on a detector, e.g. photo plate. There is an one-to-one correspondence between position on the object and the position on the detector (only valid in the transverse plane). Thus with the help of lenses relative positions on very small quantum objects can be enlarged and thus become observable. It should be noted, that a "microscope" device for magnifying the geometrical size of an atom (about 10^{-8} cm diameter) and also magnifying the relative positions of atoms in a molecule to the macroscopic size of 1 mm must have a magnification factor of more than 10^6.

Heisenberg was convinced that the position of an electron at a given time could be measured even with "ultimate" precision using the technique of such a light microscope if the wavelength of the light would be small enough to resolve subatomic structure.[1] At a "given time" means always an exposure time period in which

---

[1] Several reasons prevent that Heisenberg's so-called γ-microscope can measure the position of one selected electron inside an atom at a given time with a required resolution of 10^{-10} cm or even better: First: Since the focus of the γ-pulse is of macroscopic size (larger than 1 μm^2) the scattered photons of the γ-pulse interact with different electrons in the atom or molecule and the measurement can on principal not identify which photons were scattered on the one special electron. Second: The Compton cross sections for scattering photons with a wave length of 10^{-10} cm (or $h\nu = 1.2$ MeV) on an electron are smaller than 10^{-25} cm^2. Therefore it requires per attosecond pulse more than 10^{+19} photons in a focus of 1 μm^2 to scatter about 100 photons on this electron. Such a photon pulse carries an energy power equivalent with 1% of the total energy emission of the sun. Third a technical reason: The γ-microscope needs a high precision lens system for 1 MeV photons to magnify the 1 μm^2 focus size to make the different electrons on a macroscopic detector distinguishable.

the location of a moving electron must be considered as "frozen". Such a time period for an electron detection must be shorter than one attosecond.

Therefore, in order to obtain an image of the position of an electron with sub-atomic resolution using a γ-microscope, one would have to scatter on the **same electron** numerous photons in a one attosecond "exposure" time period (since the electron is moving with a typical velocity of 1% of the speed of light). Because these γ-scattering cross sections (Compton scattering) are of nuclear size the photon pulse intensity in one attosecond must exceed 10^{19} photons per pulse in a focus of 1 μm diameter. A further problem in such a measurement is that the experimenter has no control on which electron in the target atom or molecule the photons are scattered. Both effects make such a γ-microscope measurement physically not feasible.

Furthermore, each Compton scattering process, as mentioned above, is destructive for the electronic state, thus the electronic state changes immediately. This disturbing effect of momentum transfer to the electron and thus changing the electron's position subsequently was already realized by Bohr [19]. These arguments show that Heisenberg's γ-microscope is not suited to measure the position of an electron at a given moment.

In one attosecond exposure time because of the tiny cross sections at most one photon might be scattered on the same electron. Thus the only information the experimenter obtains with Compton scattering is the detection of only one single photon providing one momentum vector. Even if this photon momentum vector is measured with sub-atomic precision the location of the reaction can never be deduced from this one vector with a precision better than the preparation of the target position before the scattering.

In contrast, position-measurements of heavy nuclei or atoms can be performed with a γ-microscope, since the velocities of atoms or nuclei are typically a factor of 10.000 smaller. Thus, the heavy particle position can be considered as "frozen" even for an exposure time of a few femtoseconds. Such relative position measurements of heavy atoms in molecules are now routinely performed with FEL X-ray pulses [20], where a lateral position resolution of about 5 Å is achieved. A slightly better resolution of about 3 Å is achieved with CRYO-electron microscopy [21].

One may expect that when performing a multi-coincidence measurement, i.e. measuring the momentum vectors of several fragments of the same reaction with excellent resolution, one could deduce the position, where the reaction took place,

---

Heisenberg proposed also to use energetic α-particles as scattering projectiles (because of their even shorter de Broglie wave-length) for a super high-resolution microscope and estimated even a position resolution of 10^{-12} cm as possible. He wrote on page 175: "*When two very fast particles succeeding each other scatter in a very short time distance Δt on the same electron, then the distance between the positions of both collisions is Δl. From the scattering laws, which has been observed for α-particles, we can conclude, that Δl can be made as small as 10^{-12} cm, if Δt can be made sufficiently small and the α-particles fast enough.*"

For a several MeV α-particle beam this requires a relative distance of the α-particles in the beam of about 10^{-11} cm. This relative distance of the α-particles is about one thousand times smaller than the normal inter-nuclear distance in a hydrogen molecule. α-particles with a relative distance of 10^{-11} cm would repel each other with a huge Coulomb repulsion force creating huge non-controllable transverse momenta of the α-beam.

by reconstructing the intersection point of all momentum vectors. Even if the impact positions of all fragments on the detector could be measured with atomic position resolution, the momentum vectors have still a finite angular uncertainty limited again by the target preparation. Because of the macroscopic dimensions of the detection device even a tiny angular uncertainty of these vectors would spoil any precise position measurement of the reaction region within the laboratory frame.

6   Product of Precisions in Momentum and Precision in Position in a Real Measurement of a Freely Moving Single Electron

The paradigmatic demonstration experiment for the UR given in textbooks for measuring simultaneously position and momentum of an electron (wave picture) is the scattering of this "wave" on a narrow slit (first step). The scattered wave yields an interference pattern of the electron wave on a screen (second step). According to most of the textbooks position and momentum of these electrons can only be measured in such a **two step-approach**, where in the first step the position is measured by the slit width and in the second-step by the interference pattern on the screen the momentum. The electron is theoretically described by wave functions which are different before and after the slit: before passing the slit the electron is described as plane wave with well-defined momentum eigenvalue but not localized in x-position; just after the slit the electron is described as a wave packet with some distribution in position and momentum. Thus on its way to the screen the electron is in a state which is not an eigenstate of the momentum operator. In both of the two time steps the UR is fulfilled. This is a result of the fact that the two operators do not commute. Thus the wave function is disturbed and a conceptually unavoidable uncertainty in the second-step measurement (momentum measurement) is generated. From the interference structure in the transverse momentum distribution of many single-event measurements the de Broglie wave length λ and thus electron momentum p can be determined.

We will now estimate how small the product of the two precision widths $\Delta x \cdot \Delta p_x \geq \hbar$ can be made in a **single-event process** by using a modern state-of-the-art detection device. We are in particular interested in whether the product of the experimental position resolution times the experimental momentum resolution can be made smaller than $\hbar$. With today's detection technique the two-step detection scheme can be replaced for single electron detection by a quasi **one-step detection approach**, where the narrow slit is "upgraded" to a very small pixel detector, which measures position and time of impact too. Thus we consider the momentum and position at the time when a single moving electron impacts on the position-sensitive detector. One can construct detectors which can measure the impact position of the electron on the detector with a few a.u. precision $\delta xy = 2$ a.u. (see Appendix C).

In a single event this detector provides, at the instant of electron impact, also a very fast electronic signal (time resolution <50 pico-seconds) which yields precise information on the electron velocity. If furthermore the location of the interaction region and the interaction time, from where this electron is emitted, are known with macroscopic precision, one can determine the electron Time-of-Flight TOF. Knowing precisely the distance d between emission point to detector (e.g. d = 2 m and Δd = 0.1 mm) one can precisely calculate the electron velocity: Assuming the measured TOF is e.g. 456.6 ± 0.1 ns and d is 200.00 ± 0.01 cm the electron velocity is then $v_e = 2 \cdot 200.00$ cm/456.6 nanosecond $= 4.3800 \times 10^{+8}$ cm/s with an error bar of $\pm 0.025\%$. Transforming the velocity in a.u. we obtain for $v_e = p_z = 2.0021 \pm 0.001$ a.u. (p_z is the electron momentum in flight direction). In perpendicular direction the errors in momentum are $\delta p_{x, y} = (0.1$ mm/2000 mm$) \cdot 2.0021$ a.u. $\approx 10^{-4}$ a.u. Thus, in case of a single event measurement the product of the experimental error bars in the momentum and position measurement can be made $\delta p_{xy} \cdot \delta xy \approx 10^{-4}$ a.u. $\cdot$ 2 a.u. $= 2 \times 10^{-4}$ a.u. which is much smaller than $\hbar$.

One could argue, however, that the detection plus preparation is still a two-step measurement. But nevertheless in a single event the product of precisions can be made much smaller than $\hbar$. Thus, once the particle has been detected, the trajectory, that the particle has travelled on in the past, can be defined such that the product of precisions of momentum and position measurement of this freely moving single electron is not limited by $\hbar$.

7   Conclusion

We have shown that in **a single-event quantum measurement** the momenta of emitted electrons or ions can be measured with high sub-atomic precision and the limits of precision for the momentum measurement are not restricted by Heisenberg's UR if one assigns trajectories to particles that have been detected. The precision in measuring positions in a single event can never approach or being better than 1 a.u. in a single-event measurement because the two conjugate parameters position and momentum do not have the apparent physical symmetry suggested by the UR, i.e., there exists a disparity in momentum compared to position measurement. The fundamental reason is: in a single event momenta are conserved with time (i.e. they are dynamically entangled), but positions are not conserved. This fundamental difference between momentum and position measurement as function of time in a quantum reaction is also apparent from the wave description (see Appendix D). The position wave functions broaden with increasing time even during a very short single-event measurement.

For a single freely moving particle in the moment of impact on the detector momentum and also position on the detector can be simultaneously detected in a single-step approach using position-sensitive detectors combined with a time-of-flight measurement. The product of the experimental momentum resolution δp times position resolution δx on the detector can be made much smaller than $\hbar$.

Acknowledgements We are indebted to John Briggs, Tilman Sauer, Bretislav Friedrich, Reinhard Dörner, Robert Griffiths and Lothar Schmidt for many very valuable discussions and corrections to the manuscript.

Appendix A

Electrons with a very well-defined momentum $p_z = 2.000$ a.u. can be created by photoionization. The primary photon energy $h \cdot v$ is chosen to be 78.988 eV $\pm$ 0.007 eV (linearly polarized photons). These photons can singly ionize the He atom. In our example a single electron with a kinetic energy of 54.392 $\pm$ 0.007 eV and He-recoil ion with a kinetic energy of 7.4 meV are emitted back-to-back. In the He center-of-mass system both freely moving particles, electron and recoil ion, have the identical momentum with opposite direction ($p_e = -p_{rec} = 2.000$ a.u. with a precision of about 10^{-3} a.u.). The angular distribution of electrons emitted by photon ionization and recoil ions is a perfect dipole distribution.

Since the ionized He atom is not at rest in the laboratory system at the moment of ionization, one must correct the electron momentum vector for the motion of the He-atom in the Lab frame. The final electron velocity is about 5000 km/sec and the internal velocity spread of the cold super-sonic He jet is below 50 m/sec. Because the He jet is moving along the negative y direction, the correction for the electron in the z-direction can be performed with sufficient precision. Furthermore, the very small momentum kick of the incoming photon ($p_{photon} \approx 0.021$ a.u.) to the center-of-mass of the He atom changes the He velocity by only about 10 m/sec. Thus the absolute value of the electron momentum is known with about 10^{-3} a.u. precision. The direction of the final electron beam can now be defined by a collimation system (collimation in x and y direction). Using a double slit system at 2 m distance in the z-direction with slit widths of 10 microns each (in x- and y- direction) the so-collimated electron beam has an angular divergence of $\delta = 20/(2 \times 10^{+6})$ rad $= 10^{-5}$ rad. The momentum exchange with electrons inside the slits due to image-charge formation is also insignificant. Thus the momenta of the electron beam have a width $\mathbf{\Delta p_{x, y, z}} < \mathbf{10^{-3}}$ **a.u.**, i.e. each electron in this electron beam has a momentum of 2.000 a.u. in the z-direction (with a precision of about 10^{-3} a.u. in all three dimensions x, y, and z). By changing the primary photon energy, the electron beam's momentum can be varied. Similar momentum resolution results can be also obtained for photon beams and slow ion beams, too.

Appendix B

Heisenberg described also a way to measure the velocity (momentum) of an electron bound in an atom. Heisenberg's approach to measure the velocity of a bound electron was copied from classical physics where the velocity of a particle is determined from the quotient of measured distance divided by measured time difference. Heisenberg's concept was to detect by γ-photon scattering at two instants in time the electron locations (for a bound electron separated by less than 10^{-11} cm). This would require a very precise simultaneous measurement of location and time within sub-attosecond time separation. As we discussed in Sect. 4 such a measurement of position and time with the required resolution is physically impossible to perform. Furthermore, Heisenberg worried that the electron velocity (momentum) just before and just after the photon (Compton) scattering process is not the same one due to the momentum kick by the photon and thus the velocity measurement before the kick seemed to be not accessible.

In 1927, Heisenberg was not aware that one could measure nevertheless both—the electron momenta just before and after the scattering by performing **a coincidence measurement of the scattered Compton-photon and the ejected electron**.

Both electron momenta just before and just after scattering can be determined with very high resolution ($\delta p = 10^{-2}$ to 10^{-3} a.u.) due to momentum conservation during the scattering process. This is possible at high photon energies and large photon momentum transfers where the "impulse approximation" is well justified [22]. In the "impulse approximation" for Compton ionization the ejected electron is treated as a quasi-unbound electron and thus the momentum change of the whole atom by Compton ionization is rather small and the remaining ion acts only as a spectator. The coincidence measurement allows therefore for a precise determination of $P_{e\ ionized}$ which is the momentum vector of the ejected electron after the Compton scattering and $P_{e\ bound}$, which is the unknown momentum vector of the bound electron just before scattering (see vector equation in Fig. 5). When Heisenberg wrote his paper in 1927, the experimental techniques to study quantum processes were in an "archaic" state compared with today. Precise timing measurements in the nanosecond regime required for coincidence measurements were beyond imagination. The first generation of coincidence technique was just invented in 1924 by Bothe and Geiger [23], but Bothe's and Geiger's time resolution was limited to a fraction of a millisecond. Instead of using high-energetic photon impact one can perform the same kind of momentum spectroscopy of bound electrons by using very fast ion [24] or electron impact [25].

Heisenberg also considered a velocity measurement of a bound electron by using the Doppler-effect (wave length-shift) of scattered red light. Heisenberg wrote on page 177 [1]: *"The velocity of a particle can easily be defined by a measurement, if the particle velocity is constant (no acting forces). One can scatter red light on the particle und measures by the Doppler shift its velocity. The measurement becomes more precise, as longer the wave length of the light is, since then the velocity change of the particle per photon becomes smaller. The determination of position becomes accordingly more unprecise predicted by equation (1) (UR). To measure the velocity at a given moment, the Coulomb forces of the nucleus and the of the other electrons must suddenly disappear, to ensure from this moment a constant velocity, which is necessary to perform the measurement."* The Doppler-effect approach would, however, only allow the determination of the particle velocity component in the

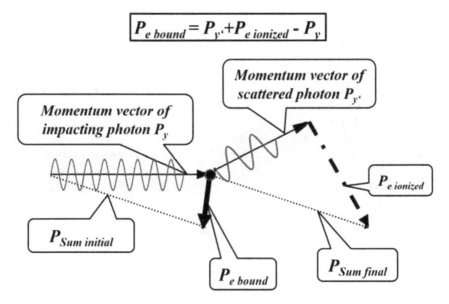

$$\boxed{P_{e\ bound} = P_{y'} + P_{e\ ionized} - P_y}$$

Fig. 5 Momentum vector diagram for Compton scattering. The dotted vectors $P_{Sum\ initial}$ and $P_{Sum\ final}$ represent the sum vectors in the initial state of impacting photon P_y (prepared and precisely known) and of bound electron $P_{e\ bound}$ (unknown) as well as in the final state of ejected electron $P_{e\ ionized}$ and of scattered photon $P_{y'}$ (both precisely measured)

direction of the incoming red light. Furthermore, the red light photon would be scattered on the whole atom and not on a single electron and would thus probe the atom velocity only.

Appendix C

Today, in principle one could build a macroscopic position-sensitive electron or ion detector with better than 10 a.u. position and 50 pico-seconds timing resolution. This detector can be a very small pixel detector or a large area position sensitive detector. Such a detector can be reassembled from two components: a commercially available position-sensitive channel-plate detector with a standard position resolution of 50 μm in x and y direction, respectively and a very thin (nanometer thickness) mask of regularly positioned holes of <10 a.u. diameter each, where electrons and ions can only be detected when they impact into such a hole (distance mask to detector surface a few nano-meter only). Then they induce in the detector an electronic signal. From this single electronic signal simultaneously position and timing information is obtained. The distance from hole to hole is 100 μm, thus the detection device is able to determine each particle impact on an absolute scale in the laboratory system for the position measurement with 10 a.u. precision. Therefore, one can detect the location of this particle impact in x and y direction with $\delta x \approx \delta y \approx 10$ a.u. resolution.

Appendix D

Time-dependent conservation laws are of fundamental importance for the implementation of a high-resolution single-event quantum measurement. This conservation is valid for the total linear momentum, for the total angular momentum and for the total energy, but not for the location. This is obvious for a particle picture, but also in the wave picture. In Fig. 6 one can see that the spatial wave function widens linearly over time, but the momentum wave function maintains its narrow width, which is determined by the preparation of the measurement. If several fragments are emitted in the reaction process, momentum conservation applies to each of the fragments on their flight to the detector. The momentum exchange with the classical electro-magnetic fields of the measuring apparatus can be determined from the classically measured trajectory of each fragment and thus corrected with high resolution.

The asymmetry as function of time for position and momentum space of the freely moving particle is also apparent in the wave approach and is due to the time propagation with $U = \exp(\Delta T \cdot E)$ with $E(p) = p^2/2$ that breaks the symmetry of position vs. momentum space.

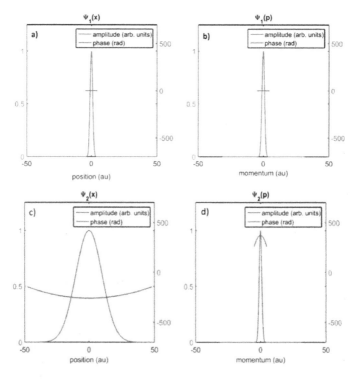

Fig. 6 Fourier limited wave function in position (**a**) and momentum (**b**) space. The phase is flat in position and momentum space and the amplitude is a Gaussian distribution. Wave function **c** from **a** after propagation for the time $\Delta T = 10$ a.u. The resulting wave function **d** in momentum space is $\Psi_2 = \Psi_1 \cdot \exp(\Delta T \cdot E)$ with $E(p) = p^2/2$. Ψ_2 has the same amplitude in momentum space as in **b** but a quadratic phase. In position space the amplitude distribution broadens compared to **a**

References

1. W. Heisenberg, Über den anschaulichen Inhalt der quantentheoretischen Kinematik und Mechanik. Zeit. f. Phys. **43**(3), 172–198 (1927)
2. W. Gerlach, O. Stern, Der experimentelle Nachweis der Richtungsquantelung im Magnetfeld. Z. Physik, **9**, 349–352 (1922); W. Gerlach, O. Stern, Über die Richtungsquantelung im Magnetfeld. Ann. Physik, **74**, 673–699 (1924)
3. https://www.scientaomicron.com/en/system-solutions/electron-spectroscopy
4. R. Dörner et al., Phys. Rep. **330**, 95 (2000) and J. Ullrich et al., Rep. Prog. Phys. **66**, 1463 (2006)
5. L.E. Ballentine, Rev. Mod. Phys. **42**(4), 358–381 (1970)
6. J.L. Park, H. Margenau, Simultaneous measurability in quantum theory. Int. J. Theor. Phys. **1**(3), 211–283 (1968); H. Margenau, Measurements in quantum mechanics. Ann. Phys. **23**, 469–485 (1963)
7. J.S. Briggs, Quantum or classical perception of Atomic Motion, Chapter 11 of these proceedings arXiv:1707.05006
8. J.S. Briggs, J.M. Feagin, New J. Phys. **18**, 033028 (2016); J.S. Briggs, J.M. Feagin, J. Phys. B At. Mol. Opt. Phys. **46**, 025202 (2013); J.M. Feagin, J.S. Briggs, J. Phys. B At. Mol. Opt. Phys. **47**, 1155202 (2014)
9. H. Kennard, Z. f. Phys. **44**, 326 (1927)
10. E.C. Kemble, Fundamental Principles of Quantum Mechanics with Elementary Applications (McGraw Hill, 1937)
11. D. Deutsch, Uncertainty in quantum measurements. Phys. Rev. Lett. **50**, 631–634 (1983)
12. M.R.H. Rudge, M.J. Seaton, Proc. Roy. Soc. London **A283**, 262 (1965)
13. W.E. Lamb, Nucl. Phys. B (Proc. Suppl.) **6**, 197–201 (1989)
14. M. Schlosshauer, Rev. Mod. Phys. **76**, 1267–1305 (2004)
15. P. Busch, P.J. Lathi, The Standard Model of Quantum Measurement Theory: History and Applications in Foundations of Physics, **26**, 7 (1996), pp. 875–893; P. Busch, T. Heinonen, P. Lathi, Heisenberg's uncertainty principle. Phys. Rep. **452**, 155–176 (2007)
16. R.B. Griffiths, What quantum measurements measure. Phys. Rev. A **96**, 032110 (2017) 14. D. Sen, The uncertainty relations in quanten mechanics. Curr. Sci. **107**(2) (2014)
17. K.R. Popper, Quantum theory and the schism in physics-from the "Postscript to the Logic of Scientific discovery page 22–23, Routledge (1989) and The Logic of Scientific Discovery page 225–226, Hutchinson
18. L.Ph.H. Schmidt, C. Goihl, D. Metz, H. Schmidt-Böcking, R. Dörner, S.Yu. Ovchinnikov, J.H. Macek, D.R. Schultz, Vortices associated with the wave function of a single electron emitted in slow ion-atom collisions. Phys. Rev. Lett. **112**, 083201 (2014); L.Ph.H. Schmidt, M. Schöffler, C. Goihl, T. Jahnke, H. Schmidt-Böcking, R. Dörner, Quasimolecular electron promotion beyond the 1 sσ and 2 pπ channels in slow collisions of He^{2+} and He. Phys. Rev. A, **94**, 052701 (2016)
19. N. Bohr, Nature **121**, 580 (1928)
20. H.N. Chapman et al., Nature **470**, 73 (2011)
21. W. Kühlbrandt, Microscopy: Cryo-EM enters a new era. eLife **3**:e03678, 4 p (2014)
22. B.K. Chatterjee, L.A. LaJohn, S.C. Roy, Investigations on compton scattering: new directions. Rad. Phys. Chem. **75**, 2165 (2006)
23. W. Bothe, H. Geiger, Ein Weg zur experimentellen Nachprüfung der Theorie von Bohr, Kramers, und Slater. Z. Phys. Band **26**, S. 44 (1924); W. Bothe, H. Geiger, Über das Wesen des Comptoneffekts; ein experimenteller Beitrag zur Theorie der Strahlung. Z. Phys. Band **32**, S. 639–663 (1925)
24. R. Moshammer, et al., Low-energy electrons and their dynamical correlation with the recoil-ions for single ionization of helium by fast, heavy-ion impact. Phys. Rev. Lett. **73** (1994) 3371; R. Moshammer, et al., The dynamics of target ionization by fast highly charged projectiles. Nucl. Instr. Meth. B **107**, 62 (1996); R. Moshammer, et al., The dynamics of target single

and double ionization induced by the virtual photon field of fast heavy ions x-ray and inner-shell processes. AIP Conf. Proc. **389**, 153 (1996); J. Ullrich, et al., Recoil ion momentum spectroscopy. J. Phys. B **30**, 2917 (1997) Topical Review

25. E. Weigold, I. McCarthy, Ion Electron Momentum Spectroscopy (Springer, 1999). ISBN 978-1-4615-477

4

High-Resolution Momentum Imaging— From Stern's Molecular Beam Method to the COLTRIMS Reaction Microscope

T. Jahnke, V. Mergel, O. Jagutzki, A. Czasch, K. Ullmann, R. Ali, V. Frohne, T. Weber, L. P. Schmidt, S. Eckart, M. Schöffler, S. Schößler, S. Voss, A. Landers, D. Fischer, M. Schulz, A. Dorn, L. Spielberger, R. Moshammer, R. Olson, M. Prior, R. Dörner, J. Ullrich, C. L. Cocke and H. Schmidt-Böcking

Abstract Multi-particle momentum imaging experiments are now capable of providing detailed information on the properties and the dynamics of quantum systems in Atomic, Molecular and Photon (AMO) physics. Historically, Otto Stern can be considered the pioneer of high-resolution momentum measurements of particles moving in a vacuum and he was the first to obtain sub-atomic unit (a.u.)

T. Jahnke · O. Jagutzki · A. Czasch · K. Ullmann · L. P. Schmidt · S. Eckart · M. Schöffler · S. Schößler · S. Voss · R. Dörner · H. Schmidt-Böcking (✉)
Institut für Kernphysik, Universität Frankfurt, 60348 Frankfurt, Germany
e-mail: hsb@atom.uni-frankfurt.de; schmidtb@atom.uni-frankfurt.de

V. Mergel
Patentconsult, 65052 Wiesbaden, Germany

O. Jagutzki · A. Czasch · K. Ullmann · S. Schößler · S. Voss · H. Schmidt-Böcking
Roentdek GmbH, 65779 Kelkheim, Germany

R. Ali
Department of Physics, The University of Jordan, Amman 11942, Jordan

V. Frohne
Department of Physics, Holy Cross College, Notre Dame, IN 46556, USA

T. Weber · M. Prior
Chemical Sciences, LBNL, Berkeley, CA 94720, USA

A. Landers
Department of Physics, Auburn University, Auburn, AL 36849, USA

D. Fischer · M. Schulz · R. Olson
Department of Physics, Missouri S&T, Rolla, MO 65409, USA

A. Dorn · R. Moshammer
MPI für Kernphysik, 69117 Heidelberg, Germany

L. Spielberger
GTZ, 65760 Eschborn, Germany

momentum resolution (Schmidt-Böcking et al. in The precision limits in a single-event quantum measurement of electron momentum and position, these proceedings [1]). A major contribution to modern experimental atomic and molecular physics was his so-called molecular beam method [2], which Stern developed and employed in his experiments. With this method he discovered several fundamental properties of atoms, molecules and nuclei [2, 3]. As corresponding particle detection techniques were lacking during his time, he was only able to observe the averaged footprints of large particle ensembles. Today it is routinely possible to measure the momenta of single particles, because of the tremendous progress in single particle detection and data acquisition electronics. A "state-of-the-art" COLTRIMS reaction microscope [4–11] can measure, for example, the momenta of several particles ejected in the same quantum process in coincidence with sub-a.u. momentum resolution. Such setups can be used to visualize the dynamics of quantum reactions and image the entangled motion of electrons inside atoms and molecules. This review will briefly summarize Stern's work and then present in longer detail the historic steps of the development of the COLTRIMS reaction microscope. Furthermore, some benchmark results are shown which initially paved the way for a broad acceptance of the COLTRIMS approach. Finally, a small selection of milestone work is presented which has been performed during the last two decades.

1 Introduction

What have Stern's Molecular Beam Method (MBM) [2] and the COLTRIMS reaction microscope (C-REMI)[1] [4–11] in common? Both methods yield a very high, sub-atomic unit (a.u.) momentum resolution for low energy particles moving in vacuum. In both approaches the high resolution is obtained because the initial momentum state of the involved quantum particles is very precisely prepared. Conceptually, there is no theoretical limitation for the achievable precision of a momentum measurement of a single particle—the precision is only limited by the design of the macroscopic apparatus [1]. Developing novel experimental detection techniques and achieving higher experimental resolution are often required for advancements in science. Already Stern's second MBM experiment, the famous Stern-Gerlach experiment, performed from 1920 to 1922 in Frankfurt, yielded for silver atoms moving in a vacuum a sub-a.u. momentum resolution in the transverse direction of about 0.1 atomic units (a.u.). Stern and Gerlach achieved this excellent momentum resolution due to a very close

J. Ullrich
PTB, 38116 Brunswick, Germany

C. L. Cocke
Department of Physics, Kansas State University, Manhattan, KS 66506, USA

[1]COLTRIMS is the abbreviation for "Cold Target Recoil Ion Momentum Spectroscopy". Another, widely employed name for this technique is "reaction microscope" (REMI). Throughout this article we will use a combination of both acronyms, i.e., C-REMI.

collimation of the atomic Ag beam [12]. That way, Stern was able to show that the Ag atoms evaporated from solid silver obeyed the Boltzmann-Maxwell velocity distribution causing a momentum broadening along the beam direction. Later in Hamburg Stern used a double gear system to chop the atomic beam, which yielded also in beam direction a quite mono-energetic beam, further improving the momentum resolution of his apparatus.

At the time Stern performed his experiments (1919–1945), detectors for the detection of individual particles did not exist. Therefore, he was only able to analyze distributions of a large ensemble of individual particles. Today, because of revolutionary developments in the recent decades, as for example, in the electronic detection techniques for low energy particles, in the target cooling, and the advances in multi-parameter data storage, the AMO experimenter can detect and obtain information on single particles and even perform so-called "complete" high-resolution measurements on atomic and molecular many-particle systems. The C-REMI approach [4–11] uses detectors that can detect the position of impact of single particles with very good position resolution (50 μm or even less) and measure the arrival time of the particles with a precision of <100 ps. From these quantities the flight times of the particles and thus their velocities are determined with—conceptually—unlimited resolution. A C-REMI setup can reach a single particle-momentum resolution of below 0.01 a.u. and it can detect all fragments emitted from an individual atomic or molecular fragmentation process in coincidence. With such properties, it has been shown in the past, that the entangled dynamics occurring during such processes can be visualized and, in special cases, relative timing resolution of 1 attosecond or better can be inferred [13]. A further important aspect of the C-REMI concept lies in the multi-parameter data handling technique employed. It provides the ability to store the raw data of each detected particle in list-mode on a computer. Thus, the experiment can be replayed during the analysis of the data applying different constraints to the data and investigating different physical aspects of the process under investigation. This advantage is common in nuclear and particle physics, but has become prominent in AMO research with the C-REMI methodologies.

2   History of Stern's Molecular Beam Method: The Technological Milestones

In 1919, when Otto Stern came back to Frankfurt he began to build his first atomic beam apparatus [3, 12] stimulated by Dunoyer's experiment [14]. Already in 1911, Louis Dunoyer had published his famous work on the generation of a so-called atomic beam in the journal Le Radium 8. He had observed that the molecules of a gas that flow from a higher pressure volume through a small aperture into a vacuum (pressure < 10^{-3} Torr) move on a straight line. The development of the molecular beam method MBM became technically possible due to the rapid improvement of vacuum techniques during World War I. Diffusion pumps were invented which enabled a vacuum of below 10^{-5} Torr. Thus for a vacuum of about 10^{-5} Torr the mean free path-length of particles moving with a velocity of about 500 to 1000 m/sec is in the order of 10 m. In such a high vacuum the experimenter can perform controlled deflection and scattering measurements with very high momentum resolution. By deflection of the particle due to an interaction with a known external force (e.g. from electric, magnetic or gravitational fields) Stern could determine atomic properties as, e.g., magnetic or electric dipole moments. The MBM allowed, furthermore, to study the ground-state properties of atoms, which were not accessible by means of photon- or electron spectroscopic methods. The deflection observed in a MBM experiment corresponds to a transverse momentum transfer. This transverse momentum transfer can be determined on an absolute scale when particle velocity and mass are known. However, in all experiments performed by Stern or his group members, beginning 1922 in Frankfurt with the famous "Stern-Gerlach-Experiment" [3, 15, 16] continuing 1923 until 1933 in Hamburg and from 1933 to 1945 in Pittsburgh only deflection angles were measured using different particle detection techniques [2, 15].

Although the Stern-Gerlach-Experiment had already demonstrated in an impressive manner what is achievable by the MBM, Stern and his colleagues continued to introduce improvements, especially during Stern's time in Hamburg. They tried to increase the sensitivity of the method and, more crucially, to further improve the momentum resolution and beam intensity.

In Frankfurt Stern used in his first experiment a heated platinum wire coated with Ag paste. Then in the Stern-Gerlach-Experiment the wire was replaced by an oven, which significantly increased the vapor pressure and thus the intensity of the atomic beam. A further increase in the beam intensity was achieved by using a slit diaphragm (see Fig. 1) instead of a small hole aperture. Since the MBM only required a high resolution in one transverse direction, the beam aperture could be made very narrow in the horizontal direction (see slit width "b" in Fig. 1) which improved the apparatus' resolution, but it could be enlarged in the other transverse direction (slit length "h" in Fig. 1) by a factor of nearly 100. Stern invented the so-called "Multiplikator" [2], where many parallel beams were created in the vertical direction, thus, de facto allowing for many measurements to be performed in parallel, without affecting the transverse momentum resolution. Stern described in [2] further efforts for improvements of the transverse momentum resolution. The path lengths r and l (see Fig. 1) were increased by about a factor of 10 compared to the setup employed for the Stern-Gerlach-Experiment and by introducing rotating gears Stern obtained also a quite well-defined longitudinal beam velocity. To be able to measure the tiny magnetic moment of the proton the magnetic deflection force had to be increased (yielding larger deflection angles) and the beams (particular H_2 and He beam) had to emitted from sources operating at the lowest possible temperature. In the last experiment performed in Hamburg before Stern's emigration in September 1933, Frisch [17] tried to observe the atom recoil momentum which is transferred when a photon is emitted or absorbed from/by an atom, which had been predicted by Albert Einstein. Frisch illuminated a sodium beam at right angle with sodium D_2 light, which caused a deflection of the atoms upon absorption of the light. Stern succeeded in his Hamburg time to improve strongly the momentum resolution thus

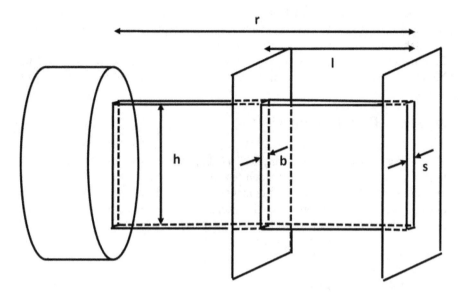

Fig. 1 Stern's method of beam intensity amplification with simultaneous improvement of the momentum resolution [2]

Frisch was able to detect this very small transverse momentum transfer of about 0.001 a.u. in this experiment, which is considered as the pioneering experiment for the Laser cooling approach. The momentum resolution obtained by Frisch is even nowadays a "state-of-the-art" benchmark achievement.

In the years 1919 to 1922 Stern employed detection techniques where a large number of the silver atoms were deposited on polished brass or glass plates in order to observe a beam spot. Later, by chemical treatments Stern was able to observe even a one-atom layer of beam deposition by silver sulfide formation observable as a black spot. In addition, the microscopic beam spot analysis (e.g. by photography) provided an excellent optical resolution in the low micron range, yet not allowing single atom counting. For beams consisting of lighter atoms or molecules, e.g., He and H_2, Stern used a different detection approach. He employed closed gas tubes with a tiny hole for beam entrance. Using very sensitive gas pressure meters he was able to obtain angle-resolved beam scattering distributions. When using H_2 beams, a further method he applied was to measure the heat increase on a metallic surface by a sensitive thermal element. Lastly, he used the Langmuir approach, as well, where the impacting atoms were ionized on a heated wire. The electric current in the wire was proportional the scattered beam intensity. The angular resolution of this method corresponded to the wire thickness [2, 16]. Stern was never able to detect single atoms or molecules.

Stern's followers, like Rabi and his scholars, used the MBM mostly for preparation of beams into selected atomic or molecular states. E.g. Townes used a Stern-Gerlach device to produce population inversion to create the first MASER device. Ramsey used two cavities with two separated oscillating fields to excite Cs atoms. In this case one could not decide in which of the cavities the atom was excited. One had to add the excitation amplitudes coherently creating sharp interference structures from which the transition frequency could be determined with excellent resolution (10^{-9}). Both Rabi's scholars were awarded the Nobel Prize in Physics (Townes in 1964 for Maser development and Ramsey in 1989 for the invention of the atomic clock) [18].

3   The C-REMI Approach

The multi-coincidence C-REMI approach [4–11] is a many-particle detection device imaging momentum space with high-resolution. The imaging is performed by measuring (in a high vacuum environment) the times-of-flight (TOF) and the positions of impact of low energy charged particles which started in a narrowly confined region in space. From these measurements the particles' trajectories inside the spectrometer volume are inferred yielding the particles' properties. This is similar to studies using the historic bubble chamber in high-energy particle physics.

In the late seventies many atomic physics groups worldwide working at accelerator laboratories investigated ionization processes in noble gas atoms induced by swift ion impact. Many research projects were dealing with the measurement of total and differential cross-sections for single and multiple ionization [19–26] (see in particular review article [26] and references therein). The resulting low-energy ions (referred to as "recoil ions" in the following) attracted interest mainly for two reasons: One research direction tried to measure the probability of ionization as function of the scattering angle by means of a projectile-recoil-ion coincidence. When measuring in thin gas targets at very small deflection angles (milli- and micro-rad) almost exclusively scattering of the projectiles from interaction with the collimation slits was observed. Therefore, to eliminate this slit-scattering problem, the projectile's deflection angle (i.e. it's very small transverse momentum) had to be measured in inverse kinematics, by measuring the transverse momentum of the recoil ion. The measurement in inverse kinematics would provide, furthermore, a tremendously improved momentum and energy loss resolution if one could bring the target atom before the collision to a nearly complete rest in the laboratory system (which was achieved later by using a super-sonic jet target [9, 11, 27] or an optical trap [28]). As an example, if in a collision of a 1 GeV/amu Uranium ion on He the projectile energy loss shall be determined, one can measure the momentum change either of the projectile or of the recoiling target atom. When detecting the projectile, the achievable resolution is limited by the properties of the preparation of the incoming beam. Even at the best existing accelerators or storage rings a relative resolution of 10^{-5} is the limit. In case of detecting the recoil ion an energy resolution of below 1 meV can be achieved, yielding a relative resolution in the energy loss of the projectile of far below 10^{-10}.

The other area of interest in the research on very low energy recoil ions, was triggered in the late seventies by the Auger-spectroscopy work of Rido Mann and coworkers [29]. He observed in high energy heavy ion-atom collisions, that (in contradiction to expectations) inner-shell Auger transitions had very narrow line widths. This indicated that the Auger electron emitting recoil ions created in these collisions stayed nearly at rest.

As first step towards developing the C-REMI approach, Charles Lewis Cocke (Kansas State University) and Horst Schmidt-Böcking (Goethe University, Frankfurt) performed together in 1979 at KSU a first test experiment to measure recoil momenta in collisions of MeV heavy ions on He atoms. In this test experiment (using a diffusive room temperature gas target, non-position-sensitive recoil detector and non-focusing recoil-ion extraction field) they measured the TOF difference between the scattered projectile and the recoil ion by performing a scattered projectile-recoil ion coincidence. The measured TOF spectrum could not be converted into absolute values of recoil ion energies, since the spectrometer could not determine the recoil ion emission angle. The results of this test experiment were therefor not published.

In the period 1982–1987 Joachim Ullrich started (as part of his Ph.D. thesis) to develop a new spectrometer approach to determine the absolute value of the transverse momentum of the recoiling target ion by measuring the TOF of the slow recoil ion emitted at 90° to the projectile beam. This development was, for the Frankfurt group, quite risky since the funding request at BMBF/GSI was officially not approved and thus one had to rely at the beginning on "self-made" equipment (e.g. detectors, spectrometers and electronic devices) [9]. Nevertheless, the project was started.[2] It was essential, that one could benefit from the experimental experience from the fields of nuclear and particle physics. C. L. Cocke and H. Schmidt-Böcking had both performed their Ph.D. research in nuclear physics and were trained in using coincidence techniques.

To accomplish the envisioned approach, novel experimental equipment, not commercially available, had to be developed. New self-made position-sensitive detectors [Micro-Channel Plate electron multipliers, MCP, with Backgammon or Wedge & Strip anodes (WSA)] for measuring recoil ions with kinetic energies between zero and several keV were developed and successfully tested. Since 1973 Schmidt-Böcking and his group had developed position-sensitive gas filled Parallel-Plate-Avalanche-Detectors PPAD for performing x-ray/electron heavy-ion coincidence measurements [30]. The work with such detectors required also experience with fast timing electronics and multi-parameter data handling and storing. With a self-made gas filled PPAD the impact time and deflection angle of the high-energy projectiles could be measured in coincidence with the recoil-ion impact time and position. This experience gave confidence that the envisioned C-REMI project was feasible. However, it took until about 1993–1995 before the first full functioning C-REMI was operating. There were moments in these years before 1990 where parts

---

[2]In the mid eighties at a small workshop on the physics at the planned TESR storage ring at MPI in Heidelberg HSB presented the perspectives on the physics with very "cold" recoiling ions. The GSI director of that time Paul Kienle heavily objected this kind of physics. Saying: we will not build a GeV accelerator to perform micro eV physics. The Frankfurt application to the BMBF to get financial support for this kind of physics and the technical developments was thus not approved, but surprisingly also not declined. Thus the Frankfurt group received support without official approval.

of the project seemed unsolvable. But Joachim Ullrich never gave up! Without his efforts and ideas C-REMI would probably not exist. Besides that, the history of C-REMI is not only a chain of recoil-ion milestone experiments performed by different groups, it is in particular the history of technological developments.

In order to finally obtain sub-a.u. momentum resolution, the target had to be prepared in a state of very small momentum spread which led to using a super-sonic jet source. A further crucial piece was specifically designed electro-magnetic spectrometer fields, that provided optimal momentum focusing with maximum detection efficiency. When, in the early nineties, the detection power of C-REMI became apparent to the atomic physics community, the Frankfurt group was ready to help other groups to build up their own C-REMI systems. Schmidt-Böcking founded in 1990 the company "Roentdek" [31] to produce the C-REMI equipment components or later even deliver complete C-REMI systems to other laboratories. Equipment was delivered to research groups worldwide by selling or in a few cases by loan. The commercial availability of C-REMI systems was essential for the propagation of the C-REMI to several new fields in Physics (single photon research, strong-field and ultrafast sciences etc.). This provided in the last two decades for many groups in AMO physics, as well as in chemistry and biology the support to perform many milestone experiments and pioneering breakthroughs. The C-REMI has enabled insight into many-particle quantum dynamics at the few attosecond scale.

3.1   The Development of C-REMI Components

In the late seventies and early eighties one of the main research activities in atomic collision physics was to measure total ionization cross sections as function of the recoil ion charge state in high energy heavy-ion rare-gas collisions using the TOF coincidence method [19–26], and to determine such cross sections differentially as function of the projectile scattering angle. These total cross sections had (for single ionization) sometimes macroscopic values [10^{+6} Mbarn $= 10^{-12}$ cm^2] and even the creation of completely ionized Ar^{18+} was possible (with a cross section about 1 Mbarn in 15.5 MeV/u U on Ar collisions [19]). Both types of experiments required a coincidence measurement between scattered projectiles and recoiling ions. Detecting the recoil ion yielded the start signal, detecting the scattered projectile with a PPAD provided the stop signal.

3.1.1   Detectors

For the detection of the high energy projectile since 1973 self-made gas filled PPADs were available, which could monitor rates up to one GHz very stably in gas flow mode (see Fig. 2) [30]. Adapted to the experimental task they measured only scattering angles by annular shaped anode structures. At that time electronics were made that enabled a simultaneous measurement of 16 scattering angles.

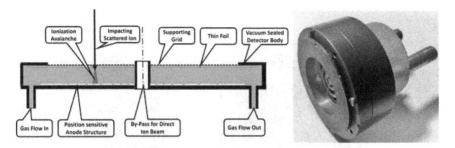

Fig. 2 Left: scheme of a PPAD. The impacting ion ejects from the entrance foil several electrons into the gas filled detector volume. In the high electric field (between entrance foil and anode structure) the electrons are accelerated and create a secondary electron avalanche which is detected as function of the anode position. This detector can have a central hole to allow the un-scattered beam to pass through. Right: A photograph of the first PPAD built in 1973 [30]. This detector had three annular anode rings and could handle rates up to 1 GHz

A position-sensitive recoil-ion detector for such low energy ions was not commercially available in the early eighties of the last century. As an initial part of a recoil-ion detector so-called micro-channel plates MCP were used [31]. The slow recoils were post-accelerated close to the MCP surface and released, upon impact on the MCP, secondary electrons that induced an avalanche inside the very narrow MCP channels. The single particle detection efficiency of standard MCP is limited by the open area ratio (e.g. how much "hole-area" is present in the total surface). Typical values are 60%. New developments of MCPs with surfaces, that look like a funnel, increase the efficiency up to 90% [33]. The position readout of the MCP was performed using a "Wedge and Strip" anode structure. Located behind the MCP, this anode structure yielded information on the position of impact of the primary particle by means of a charge partition method (see Fig. 3 [31, 32]). During the Ph.D. work of Ullrich the anode structures were fabricated as printed circuits. Prior to use, they all needed a careful restoring work by using optical microscopes. In later years such anodes were carefully printed on ceramics and did not need any initial reconditioning. A breakthrough in ion detection was achieved by Ottmar Jagutzki [33] using the delay-line approach for determining the impact positions of the particles on the detector. For such detectors the signal read-out proceeds via a delay-line structure (see Fig. 4: double-wire structure). From the arrival-time difference at both ends of the delay-line system, the position of particle impact can be determined with a resolution of better than 100 μm. The delay-line approach yields several important advantages as compared to the charge-partition method. It can handle much higher detection rates, since it does not rely on slow charge collection processes. It can detect more than one particle at (almost) the same time (i.e., being multiple-hit capable) because the induced timing signals are very short (in the range of 5–10 ns) and, lastly, the use of a "timing approach" fits perfectly to the digitized world of computers and is easy to adjust and much cheaper to build.

The first generation delay-line detectors consisted of two separate delay lines mounted at right angle. Later, Jagutzki developed a three-layer delay-line structure

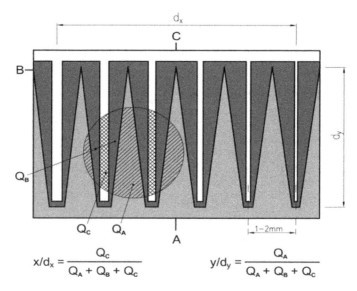

$$x/d_x = \frac{Q_c}{Q_A + Q_B + Q_c} \qquad y/d_y = \frac{Q_A}{Q_A + Q_B + Q_c}$$

Fig. 3 Scheme of the Wedge & Strip anode (WSA). If the charge cloud covers several (at least 2) pitches of three electrodes (B "wedge", C "strip" and A "meander"), measuring the relative charge portions Q_i allows to determine the centroid of the charge cloud [31, 32]

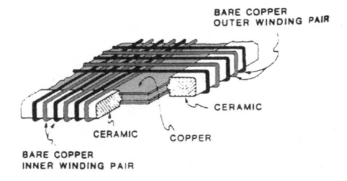

Fig. 4 The principle set-up of the delay line anode and other references therein [33, 34]

(a so-called hexanode) (see Fig. 5). The hexanode detector yields a better linearity and an improved multi-hit resolution with smaller dead-time blockade. The working diameter of these circular delay-line detectors can be as large as 120 mm diameter and recent developments target 150 mm. By using its three delay-lines the hexanode registers redundant information on each particle's position and impact time. Thus, it is possible to recover position and time information for several particles beyond the electronic dead-time limit: Even simultaneously arriving particle pairs can be detected as long as they have a minimum spatial separation of 10 mm.

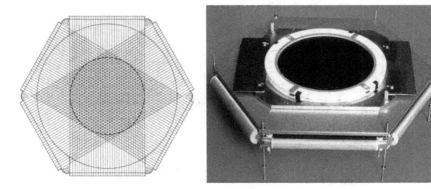

Fig. 5 Left: the Hexanode structure and right: working detector system (active area 80 mm diameter) [33–35]

3.1.2 Multi-parameter Data Handling

In the late seventies coincidence measurements, which were standard in nuclear physics, were very rarely performed in atomic physics. Thus, there was no need for fast electronics and many-parameter data handling. The electronic hardware for such measurements was quite expensive at that time and thus the comparably small groups of the atomic physics community could not afford to perform coincidence measurements where multi-parameter data had to be registered and stored. Only in nuclear and high-energy particle physics were multi-particle coincidence measurements commonly used. In order to have access to such measurement infrastructure, the Frankfurt atomic physics group, for example, performed all coincidence experiments in nuclear physics laboratories either at GSI-Darmstadt or at the MPI for Nuclear Physics in Heidelberg, where the needed electronics and data storing systems were available. The support by Ulrich Lynen [36] and Reinhold Schuch is highly acknowledged and was absolutely essential for the ongoing development of the C-REMI. Since about 1985 Ullmann [37] developed a PC based multi-parameter data storing system which was cheap and powerful enough to satisfy the needs of a two-particle or even 7 parameter coincidence measurement (implemented on an Atari ST mainstream personal computer). This development yielded a breakthrough enabling small groups to perform coincidence experiments. The inclusion of all these improvement steps took about one decade from about 1984 to 1994. The steady progress of this project was published in the annual reports of the IKF-University Frankfurt in the eighties and nineties, i.e., 1984 to 1995.

In the early years the charge signals of the WSA detectors were registered by charge-sensitive preamplifiers and subsequently amplified by standard modules. The timing-signal was created by the "Constant Fraction Discriminator" scheme [38]. These preamplifiers, constant fraction discriminator units etc. were built in the electronics workshops of the Physics Institute in Heidelberg and GSI-Darmstadt. After the foundation of Roentdek GmbH [35] several members of the Frankfurt group

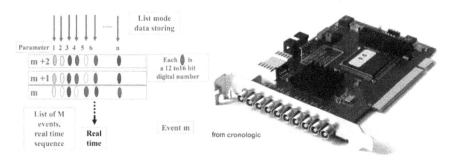

Fig. 6 Left: scheme of multi-parameter "list mode" data recording. Right: an 8-fold fast time-to-digital converter made by cronologic [39]

were employed at Roentdek and they developed their own electronic circuits. These circuits (based on modern "state-of-art" digital chips) could handle nearly unlimited numbers of parameters per event and allowed high repetition rates. Furthermore, they were inexpensive.

Simultaneously the interface between the electronic modules and the data storing PC changed from slow and expensive CAMAC to fast self-made TDC units (Time-to-digital Converters, with 25 ps timing resolution (see Fig. 6)) or even ADC units (Analog-to-Digital Converter) [39]. When using fast ADCs, the analog signal is sampled in e.g. 250 ps time-slices and its amplitude is digitized. A fast analysis program can determine several properties, as the "center" of each peak, its height or even disentangle double-peak structures (often referred to as "Camel peaks"). This development was crucial, since in case of the hexanode seven detector signals (2 for each of the three delay-line layers and one signal from the MCP to obtain the time of impact) needed to be detected for each impacting particle. Thus, a fast multi-hit recovery with very good timing resolution was needed (see Fig. 7). The present state-of-the-art C-REMI electronics including data list-mode storing can monitor coincidence rates up to several MHz.

3.1.3 Spectrometer Design

The first generation of recoil-ion spectrometers (Fig. 8) was designed to measure only total cross sections for recoil-ion production in energetic heavy-ion collisions as function of the recoil charge state and as function of the final projectile charge state [19–24]. The collimated (1) projectile beam (2) intersected with a diffusive gas jet (3). In the collision with the projectile the target atoms were multiply ionized by, e.g., pure ionization or electron capture. The projectiles (9) were deflected after the interaction with the target by a magnet (8) and detected in a position-sensitive PPAD (10). The final projectile charge states were distinguished by their bending angles behind the magnet. The low energy recoil ions were extracted by an electric field applied between plate (4) and a grid (5), which was on zero-potential. The extracted

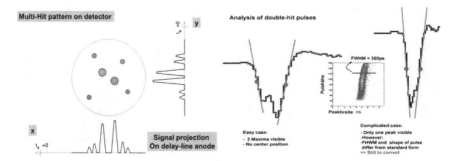

Fig. 7 Left: the circle represents the active area of the DL detector. In the event shown 6 particles impacted within about 100 ns on the detector creating electron avalanches that differ in height, which induce in the delay-line structure localized charge clouds. Each of the six ends of delay lines is connected to a fast sampling ADC. Right: the sampled multi-hit signal of one channel is shown. It is analysed later (i.e. after the actual measurement) in high detail using a PC, which allows to resolve the multi-hit pattern

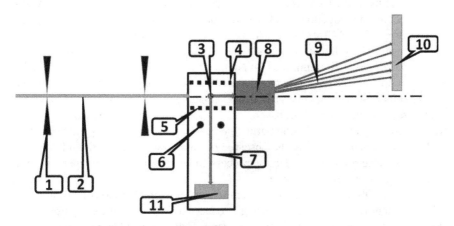

Fig. 8 Scheme of the recoil-ion deflected projectile ions (see text)

recoil ions (7) were focused on the recoil detector (11) with the help of an einzel lens (6). The recoil-ion charge state was determined by the TOF of the ions (see Fig. 9).

The second generation recoil-ion spectrometer was aimed at measuring the transverse recoil-ion momenta. This was the first working recoil-ion momentum spectrometer of the Frankfurt group. In Fig. 10 the spectrometer used by Ullrich et al. [40] is shown. The projectiles intersected with the diffusive gas target inside a field-free cylinder and were detected downstream with a PPAD. Between the inner and outer cylinder 700 V were applied to post-accelerate the recoil-ions transverse into the recoil-ion spectrometer. These accelerated ions were focused by an einzel lens. In a small magnet the recoil-ion charge states were separated and monitored by a one-dimensional position-sensitive channel-plate detector (the anode structure was only "backgammon"-like). Only such recoil ions were post-accelerated which passed

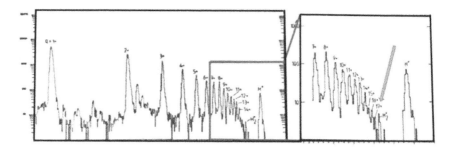

Fig. 9 Ar recoil-ion TOF spectrum after the collision with 15.5 MeV/amu U^{75+} projectiles [19]. Even the Ar^{17+} fraction is clearly visible (blue arrow). The Ar^{18+} contribution is covered by the H$_2^+$ molecular ion charge state. The Ar^{18+} production cross section is approximately 1 Mbarn

through a tiny hole in the inner cylinder. The recoil velocity (i.e., its momentum) was determined by performing a recoil-ion projectile coincidence yielding the recoil-ion TOF inside the inner (field free) cylinder. The first successful experiments investigating recoil-ion production could be performed in the mid-eighties at the Heavy-ion accelerators (UNILAC) at GSI. The first publication on the new recoil-ion momentum spectrometer with reliable small angle data appeared in Phys. Lett. A [40, 41].

For collisions of Uranium ions on Ne the transverse absolute recoil ion and the scattered projectile momenta were obtained in coincidence (see Fig. 11). The data showed that at the very small scattering angle of only μrad the sum of each recoil-ion and corresponding projectile transverse momentum did not add up to zero as expected for a two-body collision. By comparing the data to the CTMC theory of Olson et al. [42, 43], it became clear that the observed deviations were due to the influence of the emitted electron in the ionization process and due to the target temperature (the target was at room temperature), too.

From Ron Olson's calculation it became clear that internal motion of the gas target (due to its temperature) had to be strongly reduced to obtain quantitative information on the electron momenta in such measurements. Using ultra-cold targets, the method could be improved that much in resolution that electron momenta could be obtained solely by deducing them from the measured momenta of the involved ions. In his Ph.D. work Dörner et al. [44] started to build a cooled gas target. He achieved, using a static-pressure gas target, a temperature reduction down to approximately 15 K. This cooling improved the resolution, but by far not enough. In Fig. 12 the Dörner-spectrometer is shown. Inside a cooled gold-plated brass housing (connected to the head of a cryogenic pump) the cold He target is intersected by a fast, well-collimated proton beam. The He recoil ions can exit through a slit aperture towards the recoil-ion detector. Behind the slit the recoil ions are post-accelerated, focused by an einzel lens and magnetically deflected. The impact position of the recoil ions is measured by a position-sensitive MCP detector (back-gammon anode) and the TOF of the recoil ions by a coincidence with the scattered projectiles. The measurements of Reinhard Dörner showed that the expected two-body correspondence between projectile and recoil-ion transvers momentum was broken below angles of about 0.6 mrad. The first

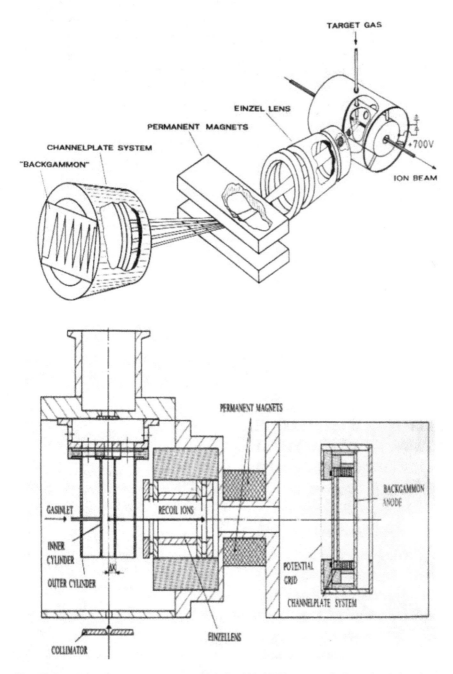

Fig. 10 Recoil-ion momentum spectrometer design [40, 41] Upper part: 3-dimensional view, lower part: cross section seen from above

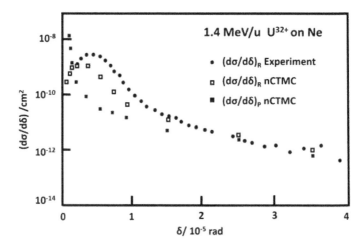

Fig. 11 These experiments were the first where—in high-energy heavy-ion rare-gas collisions ionization—probabilities at very small scattering angles <10 μrad were successfully measured. In parallel, the group of Ivan Sellin at Oak Ridge [45–47] measured mean energies of low energy recoil ions, too

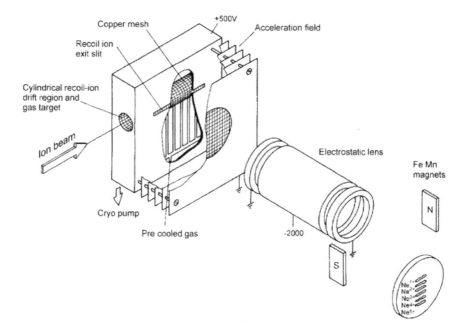

Fig. 12 Recoil-ion spectrometer of Dörner et al. [44] (see text)

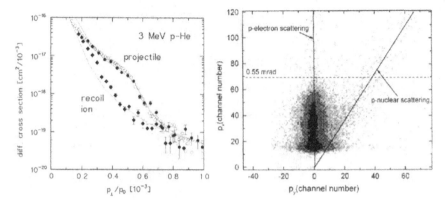

Fig. 13 Left: differential single-ionization cross sections of He in 3 MeV Proton collisions [44]. The circles show the identical data plotted versus the measured projectile transverse momentum, the diamonds versus the recoil-ion momenta. As seen in the right plot the projectile can be scattered by the He nucleus (diagonal line p-nuclear scattering) or by the He electrons (vertical line p-electron scattering). Above 0.55 mrad the projectile cannot be deflected by an electron at rest. The p-electron scattering above 0.55 mrad and the broadening below 0.55 mrad is due to the initial electron momentum (plus target temperature)

perception was: this is the principle limitation of recoil-ion momentum spectroscopy. However, with the help of CTMC calculations of Ron Olson and numerous discussions with him (Olson was a Humboldt Award fellow in Frankfurt from 1986 to 1987) it became clear, that the method (recoil-ion momentum spectroscopy) was not limited to projectile scattering angles above 10^{-5} rad, but that the method was at lower angles even sensitive to the momenta of the involved electrons if the target temperature could be decreased by several orders of magnitude. The experiment performed by Reinhard Dörner demonstrated that further target cooling would improve the momentum resolution and that it should be possible to measure the momentum exchange between nuclei and electrons with high resolution. This observation (Fig. 13) was a milestone perception towards the realization of C-REMI.

Rami Ali and Charles Lewis Cocke at KSU used the first recoil-ion extraction system [48, 49], which was time-focusing [50]. Thus the KSU group was the first to determine the Q-value (inelastic energy loss or gain) in an ion-atom collision process by measuring the longitudinal momentum component of the recoil ion [48] (i.e. parallel to the incident projectile momentum). They investigated the multiple electron capture process in 50 keV Ar^{15+} on Ar collisions and obtained a Q-value resolution of about 30 eV. Relative to the projectile kinetic energy this corresponds to a resolution just below the 10^{-3} level (see Fig. 14). The method of Q-value determination by the longitudinal recoil momentum component has been discussed before in an invited talk by Dörner et al. at the ICPEAC in Brisbane 1991 [51]. In this invited lecture, it was shown that in high-energy heavy ion collisions a relative Q-value resolution far below 10^{-6} can be obtained.

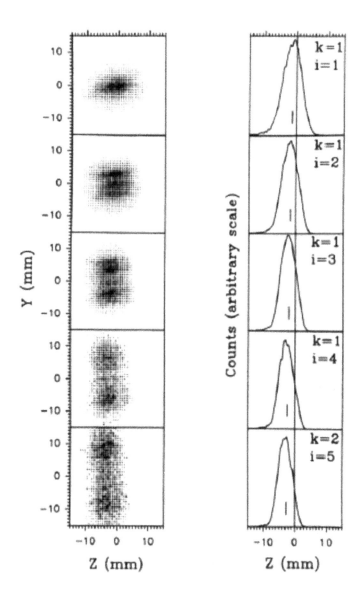

Fig. 14 Left: two-dimensional recoil-ion momentum distributions (z-abscissa: longitudinal component, y-axis: transverse component) for different projectile charge change k and recoil-ion charge state i. Right: their projections on the z-abscissa. The vertical bars indicate the center of the projections [48, 49]

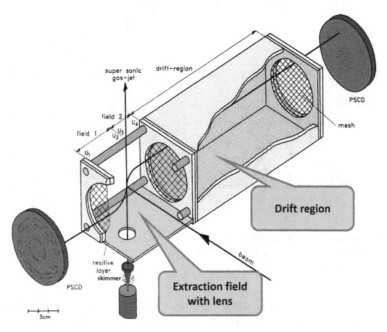

Fig. 15 Recoil-ion spectrometer used by Mergel et al. [52] with transverse extraction. The extraction field and drift zone are designed in length and field strength to obtain time focusing conditions. The detector PSCD on the right side monitors recoil ions and the detector on the left side electrons

The first breakthrough into the high-resolution domain was achieved in the "Diplomarbeit" of Mergel et al. [52] by performing an experiment with a He super-sonic jet as target and using a spectrometer design with strongly improved time-focusing properties. A further significant improvement was then achieved in the Ph.D.-thesis work of Mergel et al. [52], which used the first three-dimensionally focusing spectrometer. This spectrometer included the so called time focusing conditions and the focusing of the extension of the gas jet in the direction of the electrical field for the recoil-ion extraction. For the first time this spectrometer combined the time focusing with a two-dimensional focusing lens in the extraction field focusing the gas jet projectile reaction volume with respect to the two dimensions perpendicular to the extraction field (Fig. 15). Using this three-dimensional focusing in combination with a pre-cooled (17 K) super-sonic gas jet, a momentum resolution of 0.05 a.u. was achieved in all three dimensions, which was a breakthrough in momentum resolution of C-REMI and was the best resolution achieved at that time [52].[3]

---

[3] As Volker Mergel remembers: The three-dimensional focusing was invented in the early nineties during a night-shift performing an experiment at the tandem accelerator at KSU. In that night-shift, an experimental resolution was observed which was better than expected. These surprisingly good experimental results triggered a discussion between Charles Lewis Cocke and Volker Mergel searching for the reasons. Performing in that night-shift some calculations on the possible electric field configuration Volker Mergel could show that the reason for the improved resolution must be an inhomogeneity of the electrical field, which accidentally caused a focusing of the

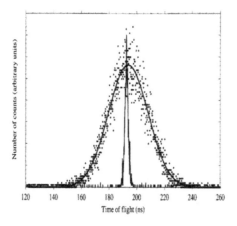 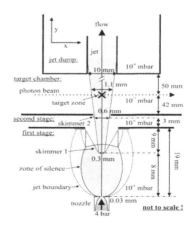

Fig. 16 TOF distribution of recoil ions emitted from a room temperature target in comparison to ions emitted from a super-sonic jet. The full line shows the thermal momentum distribution [9]. Right: Two-stage super-sonic jet [57]

In the early nineties, the Göttingen group of Udo Buck and Jan Peter Toennies [27] provided important support in constructing a super-sonic jet as target. Using this recoil technique, in the late nineties, Daniel Fischer in the group of Joachim Ullrich obtained a resolution of 10^{-6} in the Q-value measurement [11].

In the early nineties also the group of Amine Cassimi at Caen [53–56] started to use a super-sonic jet-target for recoil-ion production. Their original motivation was to build a source for intense, cold and highly-charged ions based on recoil-ion production. The effect of target cooling is visible in Fig. 16, where the recoil-momentum distribution for a room temperature and a super-sonic jet target are compared. The small transverse spread in the momentum distribution of a super-sonic jet can still be improved by collimating the super-sonic beam with skimmers (Fig. 16), thus nearly unlimited sub-atomic resolution in the transverse momentum space can be obtained.

The next very important milestone step towards today's "state-of-the-art" C-REMI system as a multi-particle momentum-imaging device was the incorporation of the magnetic field confinement of high-energy electrons (Figs. 17, 18 and 19). With this improvement, one could finally achieve a nearly 4π geometrical solid angle for the detection of even high energy electrons. Joachim Ullrich and Robert Moshammer conceived this benchmark development in the early nineties [5, 11]. It increased the multi-coincidence detection efficiency by orders of magnitude. As shown in [5, 11] the guiding magnetic field pointing in parallel to the axis of the electric extraction-field provides an unambiguous determination of the initial electron momentum, as long as the electron is not detected at flight times where its detection radius is re-approaching zero.

---

extension of the reaction volume. As a result of this discussion Volker Mergel built a new three-dimensionally focusing spectrometer as shown in Fig. 15. This technology was then also patented [DE 196 04 472 C1].

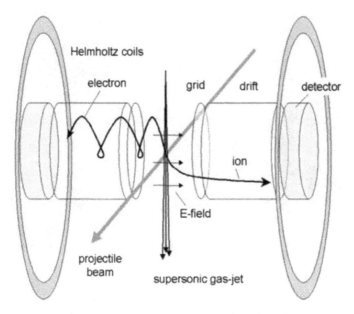

Fig. 17 The COLTRIMS reaction microscope with guiding magnetic field [11]

Fig. 18 Plotted is the TOF versus radius of electron trajectories inside the spectrometer in the presence of the magnetic field B [57]. The time between minima in the radius is the cyclotron motion period, which is used to measure the magnetic field

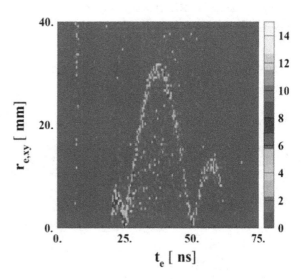

Joachim Ullrich, Robert Moshammer et al. performed several of their measurements on the physics of the recoil-ion momentum at the GSI storage ring ESR. The operation of the ESR required a wide-open spectrometer for the circulating ion beam. Therefore, they designed a spectrometer where a particle-extraction can be performed in principle in any direction (see Fig. 19), but with preferential extraction

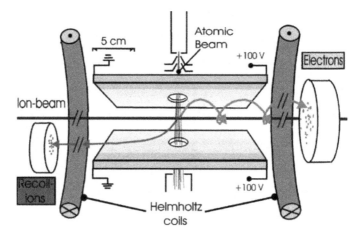

Fig. 19 The C-REMI system of Ullrich and Moshammer installed at the GSI storage ring (ESR) [5, 11]

in the longitudinal direction along the ion beam. The recoil-ion as well as the three electron detectors were positioned in time-focusing geometry [5, 11], which allowed a high Q-value resolution. Figures 20 and 21 show schematic views of modern, "state-of-the-art" C-REMI systems with transverse extraction and time-focusing. In Fig. 22 the out-side view of one of the Frankfurt C-REMI systems is shown.

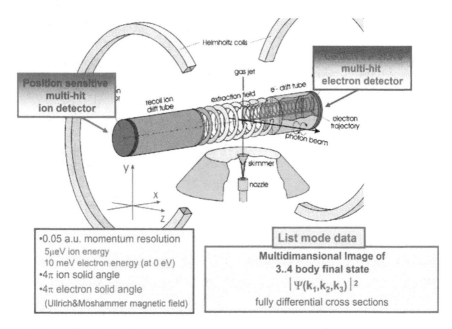

Fig. 20 Schematic view of a C-REMI system

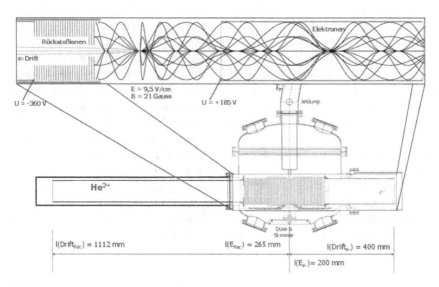

Fig. 21 Electrostatic lens-system of a C-REMI with transverse extraction showing ion and electron trajectories and time marker [58]

Fig. 22 One of the Frankfurt C-REMI systems used for MeV p on He collisions. Left: View from outside. Right: Inside the vacuum system with view on the C-REMI spectrometer [35]

4 The Early Benchmark Results

In the eighties many experts in atomic physics were skeptical that a C-REMI-like approach could actually work. The high resolution and detection efficiency of the C-REMI method was slowly recognized and acknowledged by the physics community, when in the nineties first benchmark results were obtained and published. Visiting Frankfurt, the Russian physicist Afrosimov [59] from the Joffe Institute in Leningrad (now St. Petersburg) told Schmidt-Böcking that in the fifties Russian physicist were discussing a detection method similar to a C-REMI. But they did not pursue this concept, since they did not believe that it could work because of the thermal motion of the target atoms and molecules. Using room temperature targets first angular resolved recoil-ion measurements were performed in the late seventies [60, 61]. In lectures on "Inelastic Energy-Loss Measurements in Single Collisions" Fastrup [62] discussed the advantages of recoil ion momentum spectroscopy, i.e. the method of inverse kinematics. But he also did not pursue recoil-ion momentum spectroscopy, since a target at room temperature did not allow a good momentum resolution. To our knowledge in the late seventies or early eighties no group was developing recoil-ion detection devices with larger solid angle imaging features. The required equipment like detectors, electronics, coincidence and vacuum equipment and cold target preparation methods were not available at that time to give a C-REMI a real chance of success.

4.1 Q-Value Measurements

At ICPEAC XVII in 1991 in Brisbane, the Frankfurt group presented theoretical estimates and first experimental results on the high-resolution obtainable by the cold-target recoil-ion method [51]. The prediction was that in MeV/u heavy-ion atom collisions a Q-value resolution relative to the projectile energy of below 10^{-8} and deflection angles below 10^{-8} rad could be measured. The relation between Q-value and recoil longitudinal momentum $p_{r\parallel}$ is: $Q = -(p_{r\parallel} + q/2) \cdot v_p$ [9, 11], where q is the number of electrons transferred from the target to the projectile and v_p is the projectile velocity (all values are in a.u.). The summand ($q/2 \cdot v_p$) is due to the mass transfer of electrons from the target atom at rest into the fast moving projectile system. The experimental verification of the high-resolution power of the C-REMI approach was demonstrated in the period of 1992 to 1994 by Volker Mergel [52] when he assembled the first fully working **COLTRIMS** (**CO**ld **T**arget **R**ecoil **I**on **M**omentum **S**pectrometer) system (see Fig. 15). It included a super-sonic He jet as target. The He gas was pre-cooled down to about 15 K and expanded under high pressure (>10 bar) through a nozzle of 20 μm diameter into vacuum. By the expansion process the inner temperature of the super-sonic beam decreased to a few mK.

   The beam was collimated by a skimmer (about 1 mm circular opening) to reduce the transverse momentum spread of the gas jet. Furthermore, static electric extraction fields of the C-REMI were designed to provide perfect time focusing [9, 11, 52]. Using a predecessor of the spectrometer as shown in Fig. 15 in 250 keV He^{2+} on He collisions, Mergel et al. [52] obtained an energy loss/gain resolution of 0.26 a.u. (i.e. 7 eV) by measuring the longitudinal recoil-ion momentum (Fig. 23). Relative to the kinetic energy of the impacting He^{2+} projectiles this is an energy loss resolution

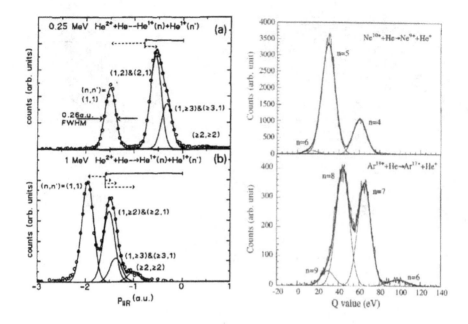

Fig. 23 Left: $p_{r\parallel}$ measurement of Mergel et al. [52]. The numbers in the brackets indicate the mean shell of the transferred electron in the initial and final state, respectively. 1 a.u. corresponds to a Q-value of 27.2 eV. Right: Q-value measurement of the GANIL group [53]. The kinetic energy of Ne^{10+} was 6.82 keV/u and 6.75 keV/u for Ar^{18+}

in the order of 10^{-5}. It is to notice that one can determine the energy loss/gain of projectiles extremely precisely without accurate knowledge of the projectile beam energy. By using the three-dimensional focusing technology of the spectrometer as shown in Fig. 15, the resolution could be improved by a factor of about 5 yielding a resolution of 0.05 a.u. in all three dimensions. The method allowed one to visualize details of electron transitions in a collision and to determine the involved electronic energy transfer with high resolution (see also [53–56, 63–65]).

4.2 Electron–Electron Contributions in the Ionization Process of Ion-Atom Collisions

In the nineties, several groups tried to separate the contributions of target-nuclei-electron (n_t-e) and electron–electron (e–e) interaction in ion-atom collisions. The electron-electron interaction can only knock out the bound electron if the mean relative velocity (projectile velocity) exceeds a certain barrier. Thus measuring the projectile ionization cross section as function of the projectile velocity the (e–e) contribution would contribute only above a certain velocity. Using the C-REMI

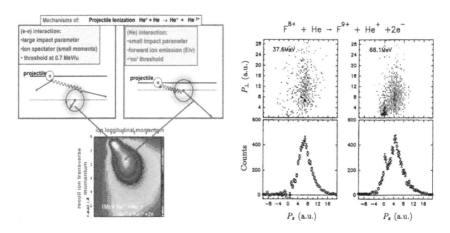

Fig. 24 Recoil-ion momentum plots for projectile ionization. Left: 1 MeV He^{1+} on He [66]. Right: 37.6 and 66.1 MeV F^{8+} on He. Density Plots and corresponding projections, the z-component is the longitudinal momentum axis [67]

approach, however, both contributions should become distinguishable in the recoil-ion momentum distribution. In the (e–e) process both involved projectile and target electrons are knocked-out and the recoil ion would act only as an observer, thus its final momentum remains at target temperature (close to zero momentum). In the (n_t-e) process the recoil ion must compensate the momentum of the electron knocked-out. In Fig. 24 (left side) the two mechanisms are explained by diagrams and the measured recoil ion momentum data for He$^+$ on He collisions are shown [66]. At this impact energy the two peaks in the distribution are clearly separated. On the right side of Fig. 24 the data are shown for F^{8+} on He [67] at two impact energies (left below the barrier, right above the barrier).

4.3 Momentum Spectroscopy in High-Energy Heavy Ion Atom Collisions

A further benchmark experiment by Moshammer et al. [68] demonstrated the high resolution power in measuring Q-values and deflection angles. In 3.6 MeV/u Ni^{24+} on He-collisions the full kinematics of the ionization process was measured by a recoil-ion electron coincidence. In Fig. 25 the sum-momentum of the electron and the He^{1+} recoil-ion is presented as function of the longitudinal momentum. More than 90% of all electrons are ejected in forward direction and their momentum is mainly balanced by the backward recoiling He^{1+} ion showing that binary projectile-electron collisions are of minor importance. The full width half maximum (FWHM) is 0.22 a.u., which corresponds to a relative projectile energy loss of $\Delta E/E_p = 3.4 \times 10^{-7}$. The obtained resolution in the transverse momentum corresponds to a resolution in

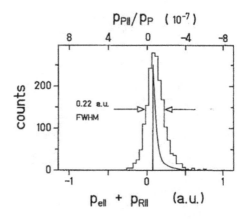

Fig. 25 Sum of electron and recoil-ion longitudinal momentum (in atomic units). Upper scale: $\Delta p_{r\parallel}$ relative to the incoming projectile momentum p_p. The histogram shows the experimental data and as a full line results from CTMC theory (normalized) [68, 69]

the deflection angle below 10^{-7} rad. In this publication for the first time the acronym COLTRIMS was defined.

In the nineties, the GSI group of Ullrich and Moshammer in cooperation with the Frankfurt group and partly with the CAEN group of Amin Cassimi explored in several research projects the mechanisms of multiple ionization of rare gas atoms in high-energy heavy ion impact. The C-REMI method allowed visualization even for GeV projectiles the energy loss at the few eV level (below 10^{-8} precision) as function of the projectile deflection angle. In Fig. 26 examples of such data are shown for He (left side) and Ne (right side) as a target [68–74]. At higher projectile velocities the momentum distribution of the ejected electrons and of the recoil ions [72] becomes more photon-like. The projectile provides by virtual photon interaction the energy for the ionization process. With higher projectile charge and slower the projectile velocity the electrons are increasingly ejected in the forward direction. In another kinematically complete benchmark experiment on single ionization of He in collisions with 100 MeV/amu C^{6+} the collaboration found small but significant discrepancies between experiment and theory which were interpreted to be the result of higher-order effects in ionization [75]. Today, the puzzling contribution of such presumed higher-order contributions remains a matter of discussion.

4.4 Single-Photon Ionization

Since 1993 the C-REMI technique also contributed strongly to the field of single-photon induced ionization processes. At HASYLAB/DESY-Hamburg and the ALS/LBNL-Berkeley first experiments with C-REMIs were performed. The C-REMI apparatus installed at Berkeley in the group of Michael Prior was mainly funded by the Max Planck Forschungspreis (200.000 DM) awarded together to Cocke and Schmidt-Böcking in 1991. Additionally, Kansas State University provided

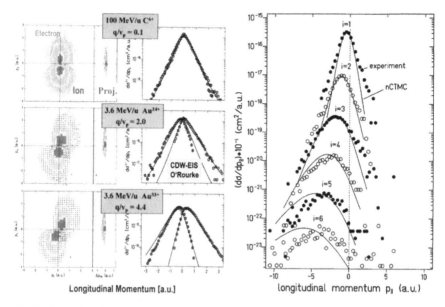

Fig. 26 Left side: electron-recoil ion momentum plots for heavy-ion impact on He measured in single event coincidence (projected on the plane of incoming projectile—recoil momentum vectors). The plotted projectile momentum vector is the sum of recoil ion and the electron momentum vectors. In the middle the projection of the projectile momentum change is plotted in a.u. [69]. Right: projections of the recoil-ion momentum in multiply ionizing 5.9 MeV/m U^{65+} on Ne is plotted (in a.u.) [69]

fellowships for Ph.D. students and Postdocs. LBNL supported the Berkeley-KSU-Frankfurt collaboration with electronic and computer equipment.

The first achievement was the measurement of the ratio R of He^{2+} to He^{1+} by single photon ionization [76]. The absolute value of this ratio was debated since the methods did not allow a reliable calibration of detection efficiencies. The C-REMI approach had the advantage that the He^{2+} and He^{1+} ions were simultaneously recorded and hence the product of photon beam intensity times target thickness and geometrical solid angle was identical, only the detection efficiencies of the channel-plate detector for He^{2+} and He^{1+} could differ.

Since the height of the detector signal was recorded for every event, too (see Fig. 27 left side), it was evident that the efficiencies for both He charge states were identical, as well. When the ratios were analyzed, however, they did not agree on absolute scale with the standard data available in the literature at that time. The data by Dörner et al. were about 30% lower than the "official" numbers published as reference values. Thus, Dörner et al. began a long search for possible unknown systematic errors in their data analysis. On a meeting at RIKEN/Tokyo in 1995 the Dörner et al. data were compared to new theoretical calculations of Tang and Shimamura. These experimental and theoretical data agreed nicely on an absolute scale within their error bars. Consequently, both were immediately published. The

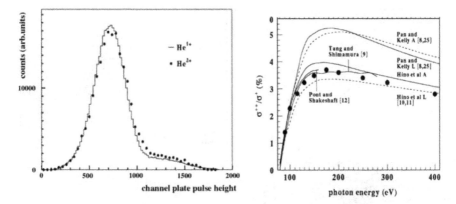

Fig. 27 Left: pulse height distribution from the channel-plate detector for He^{1+} and He^{2+}. Right: the ratio R (full circles) from Dörner et al. [76] as function of the photon energy

"photon ionization community" reacted friendly and acknowledged immediately that the standard data used so far were, for an unknown reason, increased by 40% and could be wrong. The new published data were then accepted as reliable reference.

The ratio of the total ionization cross sections of Helium occurring due to the photo-ionization and the Compton effect was another fundamentally important problem in photon physics at these times. The traditional methods of ion counting could not distinguish by which mechanism the atom was ionized. Both processes, however, differ in their recoil-ion momentum. In case of the photo effect the momentum p_e of the ejected electron is fully balanced by the recoil-ion momentum with $p_{recoil} = -p_e$ (see Fig. 28, left side) [77]. In case of the Compton effect the recoil ion acts only as a spectator and its final momentum peaks at zero (see Fig. 28, right side) [77]. Using the C-REMI approach, these different momentum distributions could quite easily be measured and separated. For photoionization one obtained, furthermore, information on two-electron correlations were the second He electron is simultaneously excited to higher n states (see Fig. 28, left side, rings of smaller electron momenta).

Correlated two-electron processes, like the double ionization of He by a single photon, were, in the nineties very hot topics in the field of photon physics performed at synchrotron machines. Pioneering, fully differential data on the subject were measured by Volker Schmidt's [79] and Alan Huetz's groups [80] by performing electron-electron coincidences. They used traditional electron spectrometers which had compared to C-REMI very small solid angles (resulting in a coincidence efficiency below 10^{-6}). The C-REMI approach has a coincidence efficiency of almost 50% and could image in quasi "one shot" the complete differential distribution (see Fig. 28, left side). Thus C-REMI revolutionized the field of double ionization processes by photon impact. Even the multi-TOF electron spectrometers of Becker and Shirley [81] did not reach the C-REMI coincidence efficiency.

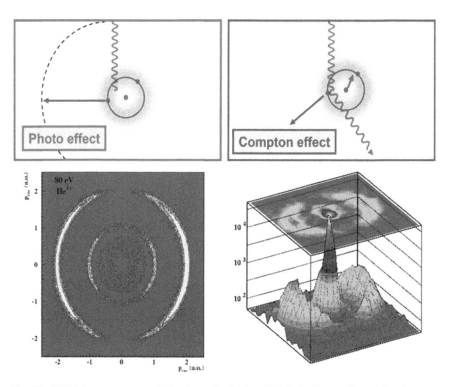

Fig. 28 C-REMI measurement of He photon ionization [77]. Left: Photo effect, right: Compton effect [78]

The first fully differential He double ionization data for circularly polarized photons were measured by Volker Mergel, Hiroshi Azuma and Matthias Achler [84] at the synchrotron machine at Tsukuba. Figure 29 shows the momentum distributions of one electron with respect to the momentum vector of the other electron for linearly polarized photons. In Fig. 30 the same plot is shown for circularly polarized photons. The asymmetric, chiral electron emission patterns are clearly visible in the distributions.

4.5   Saddle Point Ionization Mechanism in Slow Ion-Atom Collisions

In slow ion-atom collisions the mechanism of so-called saddle-point emission played an important role in the ionization process. Even when the projectile velocity was so slow that in a binary projectile nucleus-target electron collision the electron cannot be knocked out, theory predicted that the electron can be promoted to the continuum via quasi-molecular orbitals. Riding finally in the middle of the two nuclei like on a

Fig. 29 Fully differential He double ionization data for linearly polarized photons of 79 eV. The momentum distribution of one electrons is plotted with respect to the momentum vector of the other electron [82, 83]

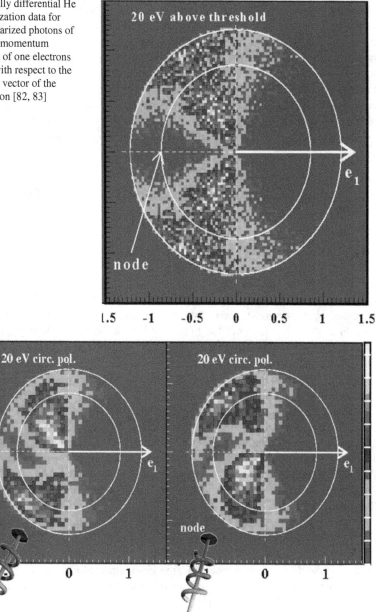

Fig. 29 Fully differential He double ionization data for linearly polarized photons of 79 eV. The momentum distribution of one electrons is plotted with respect to the momentum vector of the other electron [82, 83]

Fig. 30 Fully differential He double ionization data for circularly polarized photons of 99 eV. The momentum distribution of one electron is plotted with respect to the momentum vector of the other electron [84]

saddle, the electrons end up in the continuum in forward direction with about half the projectile velocity. Using the C-REMI approach Dörner et al. [85] investigated, at the Berkeley ECR ion source, the ionization process in slow p on He collisions, measuring the recoil-ion momentum vector in coincidence with two momentum vector components of the ejected electron. Because of the conservation of total momentum and total energy, the collision dynamics is kinematically fully defined. The surprising result was that the electrons did not ride on a saddle but their emission was kinematically steered by angular momentum conservation. The maxima of the "banana"-like electron distributions (see Fig. 31) vary in emission angle as function of projectile velocity. These shapes are centered in the nuclear collision plane.

4.6   Visualization of Virtual Contributions to the He Ground State

In 1983 Eric Horsdal Pedersen and Charles Lewis Cocke at KSU [86] and in 1986 Reinhold Schuch in Heidelberg [87] could verify, by examining the scattering angle dependence of the transfer ionization process in 7.4 MeV p + He => H° + He^{2+} + e collisions, the existence of the Thomas ionization mechanism [88]. These findings triggered great attention on the Thomas process in the whole atomic physics community. In this process the projectile nucleus can kick the bound He target electron 1 in a binary collision under 45°. On its way to the continuum electron 1 collides in a subsequent binary process with the second electron 2, thus one electron is ejected under 90° in the laboratory system and the other electron under 0°. This forward going electron is then captured by the parallel moving proton projectile resulting in He-double ionization. These billiard like two-step processes require that the projectile is deflected under the angle of $\delta_p = 0.55°$. Thus the He double ionization as function of δ_p should show a peak structure at 0.55°. Varying the projectile velocity Horsdal-Pedersen found that this maximum gets even more pronounced when the projectile velocity increases [89]. Theory, however, predicted a v_p^{-11} law [90]. Therefore, the question arose, is the peak structure at about $\delta_p = 0.55°$ really related to the Thomas process?

In the Ph.D. work of Volker Mergel the complete kinematics of the transfer ionization process in fast proton He collisions was measured by an H° and He^{2+} coincidence using a C-REMI [91, 92]. Determining the He^{2+} momentum vector and the H° transverse momentum components, the kinematics is fully controlled. In Fig. 32 the measured He^{2+} recoil-ion momentum distribution is shown for protons of 1 MeV scattered under 0.55°. Surprisingly two strong maxima appear. One at $p_{rec\parallel} = +$ 0.8 a.u. and $p_{rec\perp} = -0.5$ a.u. which coincides with the expected Thomas peak position, but the second unexpected maximum (named cKTI-p^2) at $p_{rec\parallel} = -2.8$ a.u. and $p_{rec\perp} = -1.8$ a.u. indicates there must be another, so far, unconsidered mechanism enabling transfer ionization at $\delta_p = 0.55°$. The analysis of the kinematics showed that one electron is captured at large impact parameters into the H° ground state

Fig. 31 Electron momentum distributions projected on the nuclear scattering plane for so-called "saddle point electron emission" in slow p-He collisions [85]. The momenta are plotted in relative units of the electron velocity (v_p is the projectile velocity)

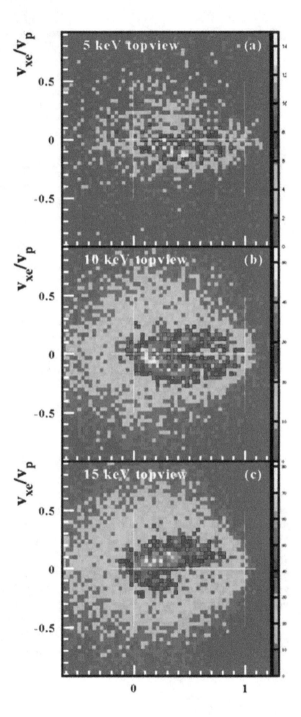

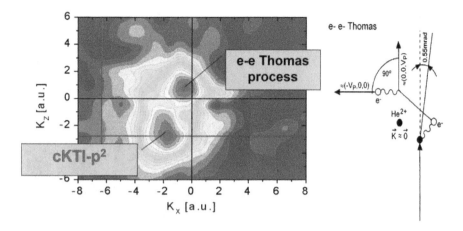

Fig. 32 He^{2+} recoil-ion momentum distribution in the nuclear scattering plane for 1 MeV on He transfer ionization process [91]. Right: The kinematics of the Thomas process

(Brinkmann Kramer mechanism) and the second He electron is emitted backward under about 135° with a momentum of approx. 3 to 4 atomic units. Mergel found the total cross section for this maximum follows a $v_p^{-7.4}$ law, thus, compared to the Thomas peak, it is the dominant transfer ionization channel at higher projectile velocities.

According to multi-configuration theory the He ground state contains a small contributions of 1–2% of the so-called pseudo-states like p^2, d^2 etc. In the p^2 pseudo-state the two electrons have opposite angular momenta and in a He atom at complete rest the target nucleus balances at any moment in a fully entangled motion the sum electron momentum to zero. If one electron in the p^2 state is captured at large impact parameters by the proton the electron 1 velocity and its direction are identical with that of the moving proton. In this moment the other He electron in the p^2 state and the nucleus must move backward. Because electron 2 is in a p pseudo-state it must enter in this moment a real continuum state. In Fig. 33 the angular distribution of the emitted electron 2 is shown in comparison to theoretical predictions. The agreement between experimental data and theory is rather good giving confidence that the presented explanation is valid. It is really surprising that such "virtual states" can be visualized in the real experimental environment. Even the kinematics at a given virtual excitation energy is visible. These tiny contributions represent an extremely small part of states contributing e.g. to the Lamb shift. However, the C-REMI approach is sensitive enough to probe the kinematics of such very small fractions of virtual states.

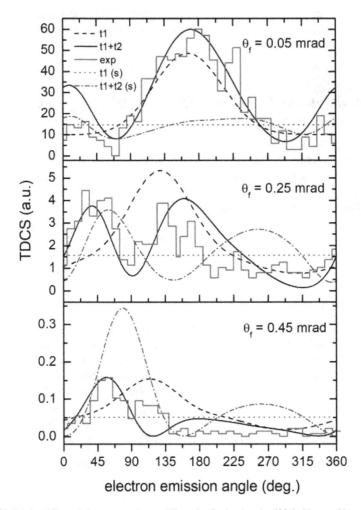

Fig. 33 Triple differential cross sections of Transfer Ionization in 630 keV p on He collisions and 20 eV kinetic energy of the electron corresponding to maximum two (Fig. 32) for three different projectile scattering angles. The black solid line is the theoretical prediction for the non-s^2 contributions. Theory and experiment are relatively normalized [93–95]

5　Milestone Discoveries

The C-REMI had grown into being an established experimental approach to study dynamics in quantum systems in Physics, Chemistry and other fields in the mid-nineties. In several hundred laboratories worldwide C-REMI systems are operating, partially commercially purchased or self-made. By using the C-REMI imaging technique many groups have produced numerous milestone discoveries. However, reference to all of these in this review paper would exceed the purpose of this article. To present all milestone results produced by the authors of this paper would also over-shoot the capacity of this review. Thus only a few of those achieved by the authors of this paper are presented here.

5.1   Multi-photon Processes—Experimental Verification of Re-Scattering Mechanism

To explain the processes underlying multiple ionization and the high double ioniza-tion probability of He and other rare gases in intense Laser pulses, Paul Corkum (1993) and Kenneth Kulander (1995) proposed the so-called re-scattering model [96]. There, emitted electrons are oscillating in the strong Laser field and are re-scattered at their parent atom. At that time period, the strong-field community did not have the proper detection device to verify experimentally this hypothesis. To visualize the dynamics of this re-scattering process one had to measure the momenta of two or more ejected electrons (and if possible of the recoil ion, too) in coincidence with high resolution. Thus, two independent collaborations, which stayed in very close contact, performed, in parallel, such coincidence experiments.

The collaborations consisted of, first, the Heidelberg group of Joachim Ullrich and Robert Moshammer, who supplied a C-REMI and joined the Laser group of Wolfgang Sandner and Horst Rottke in Berlin [97] and, second, the Frankfurt group of Reinhard Dörner and Thorsten Weber supplying the C-REMI and joined Harald Giessen in Marburg providing the Laser [98]. Presented here, in Fig. 34, are only the data of the experiment performed in Marburg [98]. For 220 femtosecond long Laser pulses of 800 nm wave length at intensities of 2.9 till 6.6×10^{14} W/cm^2 the He^{1+} and He^{2+} recoil ion momenta were simultaneously measured.

The surprising result was: the He^{1+} recoil momenta are strongly directed parallel to the Laser electric field with much smaller momenta in the transverse direction- even much less than in case of single photon ionization. The He^{2+} recoil ion momenta are in transverse direction of the Laser field similar to the He^{1+} momenta, but parallel to the field 5 to 10 times larger. In addition, they show two maxima separated by a minimum at zero. In case of single photon ionization, the recoil-ion momentum distribution reflects mainly the momentum distribution of the electron in its initial bound state, in case of double ionization by a single photon it reflects possible electron-electron correlations in the initial state. But the He^{2+} recoil momenta never exceed the He^{1+} recoil momenta by more than a factor two. Thus, in case of Laser induced double ionization only Corkum's re-scattering mechanism can explain the observation of such large He^{2+} recoil-ion momenta parallel to the Laser field. In his model, the electron can gain in the Laser field a high ponderomotive energy yielding finally a large recoil momentum. Similar work can be found in [99]. This work provided the experimental proof that the re-scattering process does explain the dynamics of the double ionization in intense Laser field and that both electrons act coherently.

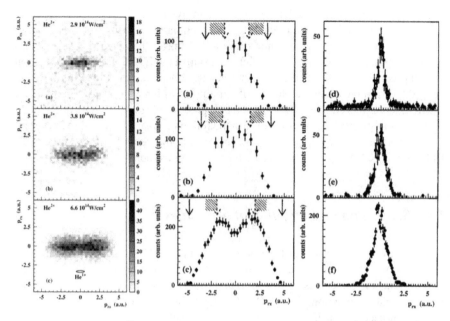

Fig. 34 Left column: He^{2+} recoil ion momentum plots for three different Laser intensities. The horizontal axis is parallel to the direction of the electric field of the Laser. The vertical axis is given by the Laser propagation. The small ellipse in (**c**) shows the half-width of the He^{1+} distribution. Middle and right column: projections of the plots of left column onto the horizontal and vertical axis. The arrows indicate the maximal momentum due to the ponderomotive energy. The dashed areas represent the in the rescattering model calculated values [98]

The two-Laser pulse pump-probe technique is well established to measure timing (delays) in the femtosecond-regime. However, using this technique in combination with the C-REMI coincidence imaging one can—despite of employing pulses of many femto-second duration—obtain timing information in the atto-second regime [100]. In 2005 the collaboration between Paul Corkums group in Ottawa and Reinhard Dörners group in Frankfurt performed such timing measurements in the attosecond regime [101] by using the Laser pump-probe scheme for ionization and for detection of the emitted low-energy electrons the C-REMI approach. The first Laser pulse aligned a nitrogen molecule and the subsequent strong probe pulse ionized the molecule. Recoil ions and electrons were detected in coincidence and from the measured recoil-ion momentum vectors the spatial alignment of the molecule in the lobaratory frame was determined. They found that both electrons did exit the molecule more likely in the same direction when the polarization of the probe pulse was parallel to the direction of the alignment. Double ionization was less probable and takes longer for the perpendicular alignment (a few hundred atto-seconds longer).

In a coincidence experiment where several momentum vectors of the emitted fragments resulting from the same reaction are detected, one can deduce from the relative angular vector directions phase differences and thus determine relative time delays. Thus a multi-coincidence momentum-imaging approach like the C-REMI method is the key to explore atto- and even zeptosecond dynamics by measuring streaking effects of Laser fields on the momenta of emitted particles. This technique allows to measure, as outlined before, time differences shorter than present Laser pump-probe technique can resolve. Ursula Kellers group at the ETH Zürich in cooperation with the Frankfurt group performed such measurements on the tunneling times in He [102]. From the observed phase shifts in the recoil ion and electron-momentum distributions it was claimed that the tunneling process takes a finite time of about 20 as, triggering strong debate on the topic in the following years. Many more important experiments have been performed in recent years in the field of ultrafast processes. References [103–112] are a few selected papers on this topic.

5.2   Single Photon Ionization of Molecules

Since Max von Laues X-ray diffraction experiment in 1912 in Munich the scattering of X-rays and electrons has been used to explore the structure of molecules. In all these studies the molecules had to be in an ordered structure (e.g. crystal) to know the molecular orientation. The C-REMI allows the study of freely moving non-oriented molecules in a gas phase. By performing multi-hit electron-ion coincidence measurements the orientation of the molecule with respect to the detection device is determined from the ionic momenta. The first successful experiments employing the idea of inferring molecular orientation from fragment emission directions were performed by Eiji Shigemasa et al. in 1995 [113] and Heiser et al. [114]. They used traditional electron spectrometers with small solid angles and had to scan the electron energy. Thus these measurements were very time consuming and gave results only for discrete angles.

The first such experiments on single photon ionization of simple molecules using a C-REMI were performed in Berkeley and in parallel in Paris. When the Advanced Light Source (ALS) started operation in 1993 Michael Prior of the LBNL in Berkeley, Charles Lewis Cocke and his group at KSU together Reinhard Dörner and Horst Schmidt-Böcking from the University Frankfurt installed a C-REMI system at the LBNL, which could be used either at the ECR source in the 88″ cyclotron building or the ALS. At the Oji-Workshop (Atomic Mol. Photoionization, September 1995) in Tsukuba, Paul Guyon and Horst Schmidt-Böcking arranged to use the C-REMI coincidence system with position-sensitive detectors to perform collaborative experiments on single photon ionization of molecules at the Paris synchrotron. Paul Guyon's group had used so far the ZEKE technique [115] to study such processes. This method had extremely small coincidence efficiency because of tiny solid angles accepted in the direction transverse of the photon beam. The Paris group provided the photon beam and gas target, the Frankfurt group the detection and data acquisition system.

First experiments on single photon ionization of simple molecules and their fragmentation by photo ionization started at Berkeley in the late nineties with Alan Landers (at that time at KSU) and Thorsten Weber (at that time in Frankfurt) being the responsible investigators. Landers et al. [116] measured the two fragment ions, their charge state and the photo electron upon C-K-shell ionization of CO in coincidence. Following the inner-shell photoionization, Auger electrons are emitted after a short delay leading to a Coulomb explosion of the molecule. Therefore, the ions' emission directions correspond to the molecular orientation at the instant of the photoionization and, from the ions' relative momenta, the kinetic energy released in the fragmentation was also obtained. For this concept to work, it is important that the delay between the fragmentation of the molecule and the initial photoionization is short compared to possible molecular rotation periods. As the photoelectron was measured in coincidence, its angular emission distribution with respect to the molecular axis was obtained.

Figure 35 shows the angular distributions of the C-K-shell photoelectrons in a polar representation, where the distance of a data point to the center of the plot represents the intensity. The double arrow with the two balls in each plot indicates the direction of the photon polarization and the molecular orientation. With the help of theory [116] details of the three-dimensional molecular potential could be deduced from such measurements. Parallel in time to the measurements by Alan Landers and Thorsten Weber et al., also the group of Anne Lafosse et al. located in Paris in cooperation with the Frankfurt group performed such measurements using C-REMI approach [117].

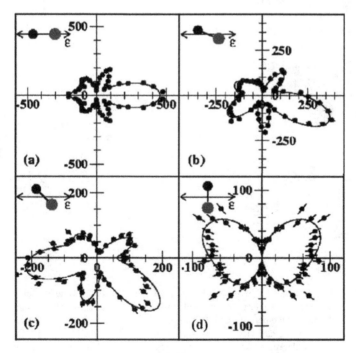

Fig. 35 Polar distribution of 10.2 eV photo-electrons in the frame of the CO molecule (small ball carbon, large ball oxygen). The solid line represents a fit to the data [118]

More photo-ionization measurements of molecules have been performed in the last two decades using the C-REMI approach (see, e.g., [118–122]). Jahnke et al. [118] performed corresponding measurements using circularly polarized photons providing first full 3-dimensional molecular frame photoelectron angular distributions, as shown in Fig. 36, left. Furthermore, they found a strong circular dichroism (CD) in the photoelectron emission (see Fig. 36, right part).

At the ALS in Berkeley Thorsten Weber and Michael Prior, in collaboration with the KSU and Frankfurt groups performed several further studies [120, 121] including the measurement of the complete photon induced fragmentation of D_2. With the support of the theory groups of Bill McCurdy in Berkeley and Fernando Martin in Madrid, fundamental information on symmetry breaking in the D_2 fragmentation processes was deduced.

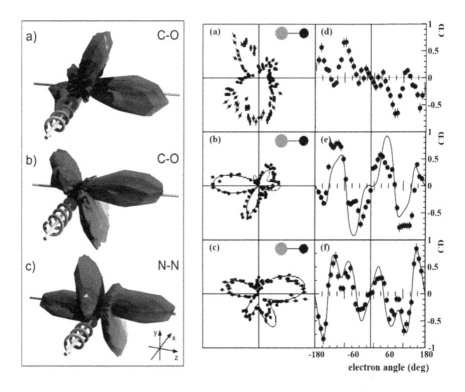

Fig. 36 Left: three-dimensional molecular-frame angular distribution of C and N 1s photoelectrons emitted from CO and N_2 molecules. The molecular orientation is indicated by the green line and the label. The handedness of the photons and their impact direction are indicated by the spirals. Right (**a, b, c**): projections of the data (left) on the plane perpendicular to the photon propagation, right (**d, e, f**): extracted circular dichroism [118]. The corresponding distributions for different molecular orientations can be found at: www.atom.uni-frankfurt.de/research/20_synchrotron/30_photon_mol ecule/20_K-shell_CO_N2/

5.3 Multi-fragment Vector Correlations in Inner Shell Single-Photon Ionization Processes of Atoms and Molecules—Dynamics of Entangled Systems

This kind of measurement approach delivers insight into two or more new fundamental aspects of atomic physics research:

1. the study of oriented very short living excited atomic and molecular ionic configurations, which can never be produced by any other preparation technique (e.g. like Laser orientation and excitation).
2. the study of dynamical entanglement in sequential cascading decay processes, exploring memory effects and dynamically induced symmetry breaking.

The multi-coincident fragment detection from a Coulomb-exploding molecule can provide insight into fundamental aspects of entangled many-particle Coulomb dynamics. In Fig. 37 the scheme of such a multistep process is indicated. From left a circular polarized photon with a well-defined energy (momentum) and angular momentum is absorbed by a two-atom molecule and creates a K vacancy thus a low energy photo electron is emitted (step 1). The electron-momentum vector (three dimensions) is measured. After a short delay, in step 2 a K-Auger electron (probably from the atom where the K vacancy was created) is ejected. The momentum vector of the Auger electron is measured too. Vacancies and excitation energy can be shared by the two atoms. In the following steps 3 to 5 more Auger electrons are emitted. From the measured Auger electron momenta, the experimenter knows the electron energy and thus the time sequence of the different steps. The delay times, however, remain unknown. Finally, with increasing degree of ionization the molecule undergoes Coulomb explosion. Measuring the ionic momenta and its final charge state the experimenter has a full dynamical control on the orientation of the molecule and on the dynamics of the fragmentation (i.e. dynamic entanglement). Finally, in this

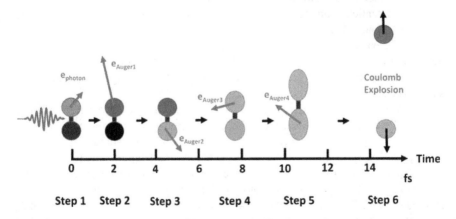

Fig. 37 Scheme of a fragmentation chain with intermediate steps 1 to 6

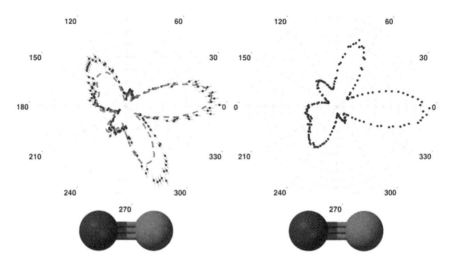

Fig. 38 Polar plot of angular photoelectron distributions in the plane perpendicular to the propagation of the photon. Left: The red dashed line represents the distribution of the right side (right handed photon) but mirrored in time. The molecular orientation is indicated by the bar-bell with Carbon on the left. Only events with a KER value >11 eV are selected, which ensures the axial recoil approximation [121]

example of fragmentation using C-REMI the experimenter has measured all together 24 momentum components and two charge states (the angular momentum vector of the photon, 5 electron momentum vectors and two momentum vectors of the ionic fragments).

Comparing this approach with the two-pulse Laser Pump & Probe technique one can "pump" (ionize) a molecule by a single high energy circular-polarized photon with subsequent photo-electron and multiple Auger-electron emission (i.e., multiple probe technique MPT). The angular momentum of the system recoil-ion and photo-electron is identical with the one of the photons (assumption: the initial state of the molecule has no angular momentum) and is therefore known by the experimenter. In this way the experimenter "pumps" the molecule by single photon absorption without destroying the dynamic entanglement of the system. Thus this MPT establishes a new field in atomic and molecular physics allowing the investigation of extremely short-lived excited molecular states.

The MPT allows one to ask whether the delayed emitted electrons have a "memory" of the earlier fragmentation steps and whether any dynamically induced symmetry breaking (in time or parity) may occur. From the measured vectors L_γ and p_{fn} (L_γ angular momentum vector of the photon and p_{fn} the momentum vector of the n-th emitted electron) one can define new dynamical coordinate systems, e.g. $L_\gamma \times p_{f1} = Z_{y1}$ and $p_{f1} \times p_{f2} = Z_{12}$ and plot the delayed electron-emission probabilities emitted in step 2,3,.. with respect to these new coordinates [121]. This new **pump & multiple-probe MPT** approach enables the investigation of fundamental dynamical processes in many-particle systems, like time or parity symmetry breaking. If time

Table 1 Vector products with respect to time and parity symmetries

Vector product	$t \to -t$	$r \to -r$
$Z = A_\gamma \times p_{ephoto}$	$Z(t) = +Z(-t)$	$Z(r) = -Z(-r)$
$Z' = (A_\gamma \times p_{ephoto}) \times p_{K\text{-Auger}}$	$Z'(t) = -Z'(-t)$	$Z'(r) = +Z'(-r)$
$S = (A_\gamma \times p_{ephoto}) \cdot n$	$S(t) = +S(-t)$	$S(r) = +S(-r)$

symmetry is broken then the distribution of the Auger electron of step 2 with respect to vector $Z_{Ly1}(+t)$ (p_{el} is the momentum vector of the photo-electron) and to vector $Z_{Ly1}(-t)$ should be asymmetrical. This vector equation shows

$$Z_{Ly1}(-t) = L_\gamma(-t)x\, p_{el}(-t) = (-)L_\gamma(+t)x(-)p_{el}(+t) = +Z_{Ly1}(t)$$

that in case of time inversion the vector does not change its sign.

In [121] for 306 eV right and left handed photons on Carbon Monoxide CO the vector correlations between the 10 eV photo electron, the K-shell Auger electron (Carbon) and the singly charged ionic fragments were measured (Fig. 38). Florian Trinter et al. [121] analyzed the coincidence data with respect to possible dynamically induced symmetry breaking. In Table 1 some "dynamical" vector products are shown with respect to time and parity symmetries.

Trinter et al. [121] have analyzed also the K-Auger electron distributions for different conditions on the momenta of the emitted photo electrons for both left and right handed photons. In Fig. 39 the K-Auger electron distributions are shown for left and right handed polarized photons with the identical conditions on the photo-electron momentum vector (in the same planes as in Fig. 38). In case of complete symmetry with respect to dynamics both distributions should have the same shape. I.e. mirroring the time the corresponding distributions did not agree within the statistical error bars. However, these preliminary measurements do not allow within their error bars any reliable conclusion on time reversal we only assert that such fundamental aspects of quantum dynamics can be explored with the C-REMI approach in these kinds of measurements.

5.4 Single Photon Induced Interatomic Coulombic Decay

Electronically excited atoms or molecules decay by photon or electron emission. More than twenty years ago Cederbaum et al. [123] predicted another very fast decay channel in loosely bound matter, where the excitation energy can be exchanged by means of a virtual photon between an excited atom and its neighboring atom. This decay channel was named "Interatomic Coulombic decay" (ICD). This process occurs in very weekly bound molecules. For example, the Ne dimer, which is a prototype system for ICD, is bound by the van der Waals forces with a binding energy of 2 meV at an inter-nuclear distance of 3.4 Å. First experimental evidence

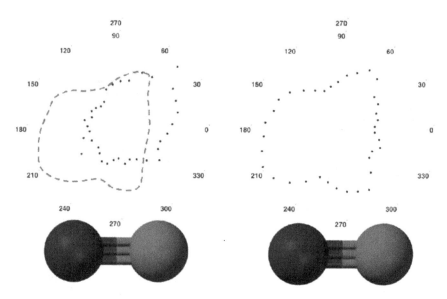

Fig. 39 Polar plot of angular K-Auger-electron distributions in the plane perpendicular to the propagation of the left and right handed circular polarized photons. Left: The red dashed line represents the distribution of the right side (right handed) [121]

for the existence of ICD was reported by observation of slow electrons emitted in large photon excited Ne clusters by Marburger et al. [124]. ICD in Ne dimers can, however, be unambiguously identified by coincident detection of two Ne^{1+} fragments and the low-energy ICD electron. To yield a unique fingerprint of this ICD process, Till Jahnke and Achim Czasch have performed at BESSY II in Berlin a corresponding multi-fragment coincidence experiment using the C-REMI approach [125]. The photon energy was chosen such that only a 2s electron in one Ne atom could be ejected, but a subsequent Auger transition in the same ionized atom was energetically not possible. As ICD occurs, the excitation energy is transferred to the other atom of the dimer causing the ionization of its outer shell. The energy released in the process is shared by the ICD electron and fragment ions. Therefore, the total sum of the kinetic energies is fixed and can be used for an unambiguous identification of the ICD process. A scheme of the ICD process is shown in Fig. 40. The quantitative values of the shared energies are plotted in a two-dimensional plot, KER energy versus electron energy in Fig. 41. The ICD feature forms a diagonal line depicting the constant energy sum, as predicted for the ICD process. In [126–131] more recent work on the ICD process is presented.

The experimental proof and verification of the existence of the ICD process was only possible by the coincident detection of all charged fragments occurring in the process. Most hydrogen or van der Waals bound systems, most prominently liquid water, will often release or transfer energy via the ICD channel. A recent, comprehensive review on ICD can be found in [132].

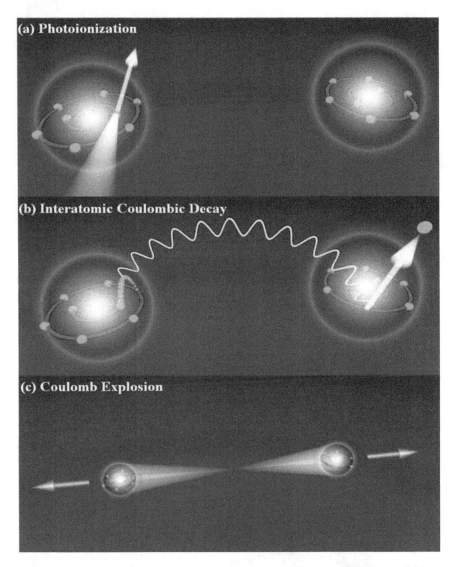

Fig. 40 Scheme of the ICD process. **a** Photoionization with ejection of a 2s photoelectron; **b** Virtual photon transfer from the ionized atom to its neutral partner atom yielding the emission of 2p electron from the partner atom; **c** Coulomb explosion of the doubly charged dimer [125]

In 2013 Trinter et al. [131] have investigated the ICD process in van der Waals-bound HeNe molecules. Najjari et al. [133] predicted that in such molecules one of the atoms can act as a very efficient antenna to absorb photons. In case of HeNe, the ionization cross section is strongly enhanced (by a factor of 60) if the photons can first interact with the He atom. It absorbs the photon and in an ICD-process the energy is transferred to the neighboring Ne atom which is then ejecting an electron.

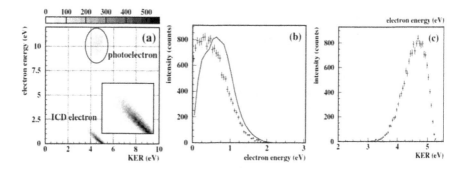

Fig. 41 Left: Kinetic energy release KER of the Ne ions versus the energy of photo electron and ICD electron. Right: Projections of electron and KER value distributions [127]

In Fig. 42 (left side) the different steps of this process are shown. The measurement was performed with a C-REMI system detecting the emitted electron and ion in coincidence. Florian Trinter et al. have experimentally verified that a single atom can act as a highly efficient antenna to absorb energy from a photon field and transfer the energy to a neighboring receiver atom within a few hundreds of femtoseconds. The resolved vibrational states of the resonance provided a benchmark for future calculations of the underlying energy transfer mechanism of ICD.

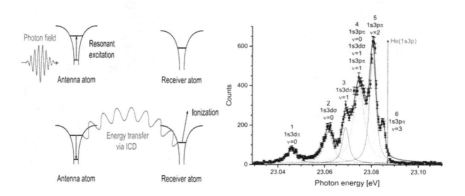

Fig. 42 Left: Scheme of absorption and decay steps. The photon coming from left is absorbed by the He atom, which is resonantly excited into the 1s, 3p state. Before it can decay by photon emission the excitation energy is transferred via resonant ICD to the neutral Ne atom, leading to its ionization. Right: the photon energy was scanned over the range of the He resonance below the actual ionization threshold. The vibrational states of the molecule can be nicely resolved (see theory [135])

5.5   Core-Hole Localization

Each atom or molecule represents one unified dynamical quantum state with a well-defined total energy, where all electrons together with the nuclei form by spin-orbit coupling one state with well-defined angular momentum and exactly ZERO total momentum in its own center-of-mass system—strictly conserved over varying time. Each atom or molecule is not a sum of single particle states, the experimenter cannot number and distinguish each electron e.g. as a specific K-shell or L-shell electron, which can be knocked-off to the continuum thus allowing an initial state localization of the ejected electron. One can only create an ionized atom/molecule in an excited new energy state with a K-shell or L-shell vacancy. In case of an inner-shell hole this vacancy may be localized for an extremely short time near the nucleus of one atom. In 2008 Schöffler et al. [134] have been able to explore this open problem by investigating the symmetry in the angular emission distributions of photo- and Auger electrons emitted from molecular N_2. In their experiment, they measured the photo- and the Auger electron, as well as the emitted ionic fragments in coincidence. The emitted electrons yielded de facto an ultra-fast probe of the shortly existing possible asymmetry of the electronic potential near both nuclei. Early theoretical calculations [135] suggested that even fully symmetric molecules consisted of asymmetric contributions in their ground-state in case of core-hole localization. This work resolved a decade of debate on possible core-hole localization with several experiments proving its existence and others concluding that core-holes are fully delocalized. The C-REMI work by Markus Schöffler et al. demonstrated, that the question of core-hole localization or delocalization remains not fully answered.

It is not only the core-hole (or the corresponding photoelectron) that needs to be considered, but the whole molecule as such. The emitted photoelectron (and thus the core-hole) forms an entangled state with the Auger electron and the fragment ions of the molecule. Depending on the properties of the entangled partners, the properties of the core-hole changes, as well. It was shown in [134] that fingerprints of a localized core-hole can be observed, if the Auger electron resides in a superposition of *gerade* and *ungerade* states, and inversely, the core-hole is delocalized if the corresponding Auger electron can be attributed to a distinct *gerade* or *ungerade* configuration. Figure 43 depicts the photoelectron angular emission distribution in the molecular frame. Panel A shows a symmetric distribution averaging over all emitted Auger electrons. The distribution in Panel B becomes asymmetric (depicting localization) as a gate on distinct Auger electron emission directions is applied (thus selecting a *gerade/ungerade*-superposition).

5.6   Efimov State of the He Trimer

Since more than hundred years, long-range van der Waal forces have attracted great interest in molecular physics. In their origin they differ from the Coulombic and the

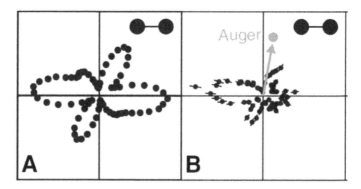

Fig. 43 Angular distribution for of 9 eV photo-electrons emitted from the K shell N_2. The circular polarized photons had an energy of 419 eV. The propagation of the photons is perpendicular to the plotted distributions. The molecule orientation is indicated by the bar-bell. A: Integrated over all Auger-electrons, B: Photon electron distribution coincident with a specific Auger electron emission direction (green arrow) [134]

covalent bonding force and are created by dynamical correlation (or better dynamical entanglement), Van der Waal forces can create bonding at huge inter-nuclear distance. Efimov [136] predicted in the late 60-ties of the last century a universal three-body state which exists as any dominating two-body force vanishes. Such bound three-body states have been termed since then "Efimov-states". It has been predicted that at very low temperature an excited He trimer molecule may form an Efimov state with several hundred Ångström inter-nuclear distance between its atomic constituents. Already the He dimer is one of the largest, naturally occurring system (exceeding by far 100 Å) with a binding energy of only a few hundred neV. It was discovered in 1994 by Schöllkopf and Toennies [137] by matter-wave-diffraction and analysis of the observed interference structures.

Starting from initial work on He dimers [138], Kunitski et al. [139] succeeded in 2015 to produce, identify He trimers in an Efimov state and measure the vibrational wave-function of the ^{4}He Efimov-trimer. They prepared the excited state by employing the matter-wave diffraction technique of Schöllkopf and Toennies [137] and multiply-ionized the trimers with a short, highly intense Laser pulse. The rapid ionization yielded a Coulomb explosion of the trimer and—using the C-REMI approach—the momenta of the ionic fragments were measured in coincidence. From the measured momenta the spatial structure was determined, which is shown in Fig. 44. The agreement between experiment and theory is very good. The Efimov trimer consists in principle of a He dimer with the third He atom orbiting at even further distance. This experiment has proven that C-REMI is able to clearly identify even very rare events in the presence of other hugely dominating processes or background, due to the coincident detection of all fragments with the precise measurement of momenta.

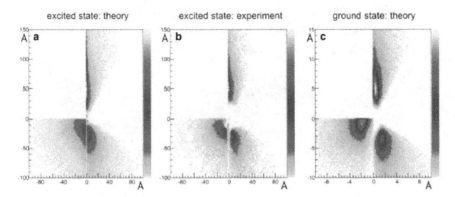

Fig. 44 Structure of the He trimer. **a** The structure predicted by theory and **b** the measured one for the excited Efimov state [139, 140]. **c** for comparison, the ground state structure as predicted by theory. Notice the factor 10 difference in the size

5.7 Imaging of Structural Chirality

Many pharmaceutical drugs have a chiral structure. Since the "Contergan" case [141] in 1961 it became clear that the purity of drugs is a crucial condition for their application. One handedness is constitutional and the opposite handedness can be noxious. Even a very small impurity of the wrong handedness can be very dangerous. Thus it would be of great help if one can recognize for each molecule whether it has the proper chirality. A C-REMI can analyze molecules in the gas phase (and in the future eventually drugs) and decide practically with 100% certainty which handedness is present. Martin Pitzer from the Frankfurt group together with the chemistry group of Robert Berger in Marburg investigated the single-photon (710 eV) and strong-field induced complete fragmentation process of chiral molecules as, for example, CHBrClF and detected the five ionic fragments in coincidence [142, 143]. The molecules are randomly oriented in the gas phase, but as pointed out before, the coincident detection of ionic fragments allows for a determination of their orientation on a single molecule basis. Moreover, when investigating larger molecules, even the molecular structure can be reconstructed from the momentum measurement. As an example, the distribution of the measured momentum vectors is shown (after multiple ionization of CHBrClF using a fs-Laser) in Fig. 45. The Carbon ion is marked by the black sphere, the H ion by the white dots, the F ion by the green dots, the Cl ion by the yellow dots and the Bromine ions by the red dots. The multiple coincidence condition of 4 or five fragments reduces the background nearly to zero and allows to distinguish molecules of different handedness from a racemat, i.e., the experimenter can extract for each ionization event the handedness of the molecule. In Fig. 46 this unambiguous identification of the handedness by using C-REMI becomes obvious. Here the data are plotted as function of the chirality parameter.

$$\cos \Theta_{F(Cl \times Br)} = \boldsymbol{p}_F \cdot \left(\boldsymbol{p}_{CL} \times \boldsymbol{p}_{Br} \right) / \left(|\boldsymbol{p}_F| \cdot |\boldsymbol{p}_{CL} \times \boldsymbol{p}_{Br}| \right) \ [137].$$

Fig. 45 Momentum vector distribution of ionic fragments of the chiral CHBrClF molecule after Laser ionization. Left: Left handedness, right: right handedness [142] (see text above)

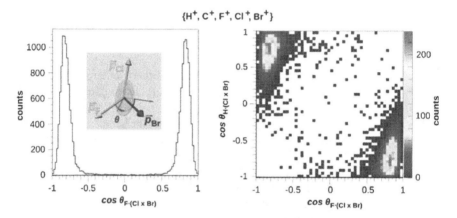

Fig. 46 Measured handedness distribution as function of the chirality parameter $\cos \Theta_{F(Cl \times Br)} = p_F \cdot (p_{CL} \times p_{Br})/(|p_F| \cdot |p_{CL} \times p_{Br}|)$ [142]

5.8 Spatial Imaging of the H_2 Vibrational Wave Function

Dependent on the gas temperature, molecules in the gas phase undergo repetitive collisions with neighboring molecules. This leads to excitation of vibrational or rotational states. The exploration of this intra-molecular motion with traditional x-ray or electron diffraction methods is complicated, as the method is not very sensitive to such features and yields the mean averaged spatial structure. Using Coulomb explosion imaging methods [143] with subsequent coincident measurement of all momenta of the ejected fragments, however, can yield information on this intra-molecular motion (i.e., the vibrational wave function of the nuclei) with high resolution. In Frankfurt Schmidt et al. [144] have investigated the vibrational states of excited H_2 molecules. 2.5 keV H_2^+ ions produced in a Penning ion source collided with a very cold super-sonic jet He beam and were neutralized by capturing one electron into the

Fig. 47 Calculated energy levels as a function of the inter-nuclear distance R for H_2 molecules depicting the concept of the "reflection approximation"

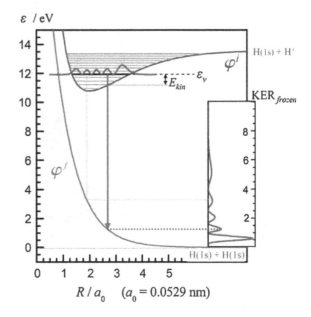

different vibrational states. Using the C-REMI approach the two neutralized H fragments were detected in forward direction by a multi-hit capable time- and position-sensitive detector and the He^+ ion was detected perpendicularly to the ion beam with a C-REMI system. Since the momenta of all three fragments were measured with high resolution (<0.04 a.u. => 3 micro eV) the kinetic energy release KER and the electronic excitation energy (different vibrational states) could be cleanly determined. From the measured H momenta, the H_2^+ inter-nuclear distance was inferred using the reflection approximation (see Fig. 47). The experimental density plot of vibrational states as function of the inter-nuclear distance and electronic excitation energy is shown in Fig. 48. The reflection methods yield slightly different results in case of approximating the nuclei as "frozen" or "moving". This difference becomes obvious from Fig. 49, when the data are analyzed for both reflection methods (green dots: frozen nuclei; red circles: moving nuclei). The solid line represents a mean value of both reflection methods.

5.9 Visualization of Directional Quantization of Quasi-Molecular Orbitals in Slow Ion-Atom Collisions

"Space quantization" or more appropriate "Directional quantization" (Richtungsquantelung) of atomic states in the presence of an outer magnetic field is known since 1916 when it was proposed by Debye and Sommerfeld [145] and its verification in 1922 in the Stern-Gerlach experiment [14]. The existence of such a directional

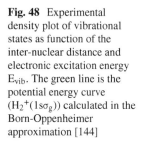

Fig. 48 Experimental density plot of vibrational states as function of the inter-nuclear distance and electronic excitation energy E_{vib}. The green line is the potential energy curve $(H_2^+(1s\sigma_g))$ calculated in the Born-Oppenheimer approximation [144]

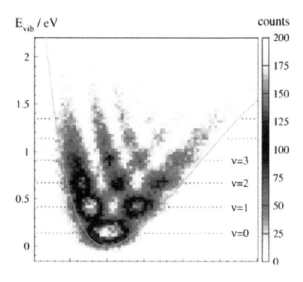

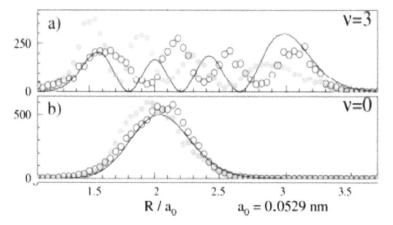

Fig. 49 Distribution of vibrational states as function of the inter-nuclear distance R, where R is calculated for the frozen and moving nuclei reflection methods (green dots: frozen nuclei; red circles: moving nuclei). The solid line represents a mean value of both reflection methods [144]

quantization also in electric field was already indirectly seen in the Stark-effect. The existence of a directional quantization of electronic quasi-molecular states was recently nicely explored by Lothar Schmidt in Frankfurt [146] where he measured the electron emission in slow 10 keV He^{2+} + He → He^+ + He^{2+} + e transfer ionization processes. By measuring all three emitted charged fragments in coincidence, the electron emission pattern with respect to the nuclear scattering plane were visualized. In this slow collision process inner-shell quasi-molecular orbitals are formed which are oriented in angular momentum (directional quantization) with respect to the nuclear collision system.

Averaging over all orientations of the nuclear scattering plane, the electron emission pattern does not show any sign of directional quantization, only when for each event the orientation of the nuclear plane is measured. In Fig. 50 the distribution of the emitted electron projected on the nuclear collision plane is shown (a experiment, c theory). The discrete structure corresponds to discrete angular momentum states. The abscissa and ordinate are given in units of the ion velocity v_p. A detailed discussion of this structure is given in [146]. In Fig. 50b, d the projections perpendicular to the nuclear scattering plane are presented. The comparison between experiment and theory shows perfect agreement and proves that in any quantum measurement where the experimenter is sensitive to angular momentum the quantum system reveals the principle existence of directional quantization, i.e. the ordering concept of dynamics in quantum systems.

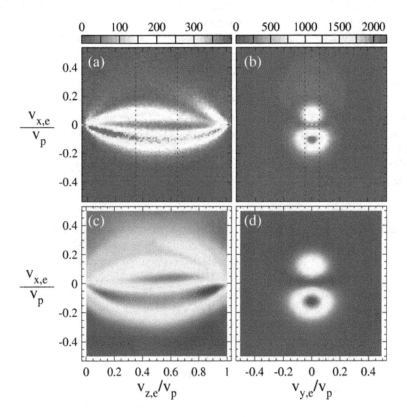

Fig. 50 Electron momentum plots in 10 keV $He^{2+} + He \rightarrow He^+ + He^{2+} + e$ transfer ionization processes [146]. The abscissa and the ordinate are in units of the projectile velocity v_p, i.e. the electron momenta, panels **a** and **b** are experimental data and **c** and **d** theoretical predictions. Panels **a** and **d** depict the projections on the nuclear scattering plane an **b** and **d** perpendicular to it

5.10   Time-Resolving Studies Employing Coincidence Detection Techniques

In the recent past it has been demonstrated, that time-resolving experiments are possible without having a projectile source with corresponding timing properties, as, for example, in a laser pump-probe scheme. In some cases, the temporal evolution on atomic or molecular time scales can be deduced from other information obtained from the coincident detection of ions and electrons. This subsection will provide three recent examples of such studies.

Interatomic Coulombic Decay (ICD) has been a subject of large interest, as described in section V.d. Its efficiency (and thus the lifetime of IC-decaying states) is strongly linked to the inter-nuclear distance between the participating entities. As typical ICD lifetimes are in the range of a few tens of femtoseconds to picoseconds, the excited compound, that will undergo ICD, will exhibit changes of its geometry prior to the decay. These nuclear dynamics triggered strong interest in performing time-resolved measurements of ICD during the last decade, because—as mentioned above—the nuclear motion alters dynamically the electronic decay probability, making ICD a prototype process for distinct non-exponential decay behavior. A molecular movie of the nuclear motion during ICD in helium dimers has been obtained in 2013 by Trinter and coworkers [147]. They used a synchrotron source for triggering ICD in He_2, which has obviously no timing properties, that allow for a direction determination of single event ICD lifetimes (typical synchrotron light pulses have a duration of approx. 100 ps). Accordingly, Trinter et al. introduced a novel approach to extract the decay time of single ICD events from their coincidence measurement. By the so-called "PCI-streaking" the decay time is encoded in the photoelectron kinetic energy. PCI (Post Collision Interaction) is an effect studied in detail already since the 1970-ties [148]. Adopted to the scheme of ICD, the following process takes place: a low energy photoelectron is emitted from a dimer creating the IC-decaying state. As ICD occurs, an ICD electron (in the case of He_2 of approx. 10 eV kinetic energy) is released. If the photoelectron has been chosen sufficiently slow (by selecting an appropriate photon energy from the synchrotron light source) the ICD electron will overtake the photoelectron which causes a change of the effective potential the photoelectron is emerging from, i.e., the potential changes from effectively singly charged to doubly charged. The more attractive potential will decelerate the photoelectron, and the amount of deceleration depends on the emission time of the ICD electron. Thus, by performing a high resolution measurement of the photoelectron momenta and the two ions created in process, the decay time can be inferred from the photoelectron energy and the inter-nuclear distance of the two atoms of the dimer from the ions' kinetic energy release. Employing this approach, Trinter et al. were able to create snap shots of the nuclear motion during ICD covering the first picosecond after the excitation.

Similarly, but using a different approach in detail, Sann et al. showed, how an electronic orbital transforms from being *molecular* to *atomic* upon dissociation of a molecule [110]. A resonant excitation of a HCl molecule triggered its (ultrafast) dissociation [149]. During the dissociation an Auger electron is emitted. Depending

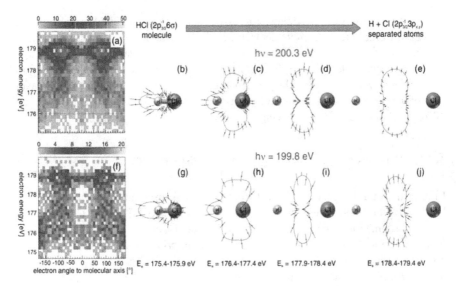

Fig. 51 **b–e** and **g–j** show the transformation of the molecular frame angular distributions of Auger electrons emitted during ultrafast dissociation of HCl. The MFPAD shows initially molecular features (left) and becomes atomic for larger inter-nuclear separations (right). The figure has been taken from [110]

on the emission time, the Auger electron is either emitted from the still intact HCl molecule, an intermediate state or—at later times—from the Cl atom. Sann and coworkers investigated the molecular frame angular distributions of the Auger electron for different inter-nuclear distances during the dissociation. The inter-nuclear distance has been inferred from the energy of the ion measured in coincidence with the Auger electron, providing (as the overall dissociation and decay process typically occurs within 5 fs) information on the timing, as well. The molecular frame angular distributions changed during the dissociation from showing signatures of a molecular orbital to an atomic distribution as shown in Fig. 51.

Very recently, Grundmann et al. investigated the following question employing a multi-particle coincidence approach [13]: Is an electron emitted simultaneously from all across a molecular orbital as it is released by photoionization, or is it first released from that portion of its orbital that is "illuminated first" by the photon? A H_2 molecule has been used as prototype test bench to answer this question. Photoemission from a homo-nuclear molecule can be—due to the two-center nature of the molecule—intuitively regarded as a microscopic analog to scattering at a classical double slit. The photoelectron wave is emitted as a superposition from the "left" and the "right" atom of the molecule, which, indeed, causes Young-type interference patterns in the molecular frame angular emission distribution of the electron. Grundmann and coworkers showed, that the molecular frame angular emission distribution changes subtly if the molecule is oriented along the photon propagation direction or perpendicular to it during the photoionization process. Within the double slit picture these changes are understandable: if the molecule is oriented perpendicular to the photon direction, the photon arrives at both nuclei of the molecule at the same time. However, if it oriented in parallel to the propagation direction, one of the atom is hit prior to the second one. This delay in the arrival time can be modelled as a phase shift of the one of the emerging photoelectron waves, or, in the double slit picture, a phase shift in one of the two slits, which causes a measureable displacement of the double slit interference pattern. From this displacement a birth time delay of approx. 250 zeptoseconds was resolved in the experiment, which nicely corresponds to the travel time of the photon along the molecule. The sensitivity of this approach is below a few 10 zeptoseconds, despite employing synchrotron light pulses of >100 ps duration.

5.11 *Proposed Experiments in Neutrino Physics*

In an article published in 1994 in Comments on Atomic and Molecular Physics Ullrich et al. [150] presented future perspectives of the C-REMI technique. One exotic one is worth mentioning here: With the C-REMI approach, i.e. measuring with ultra-high resolution the momenta, one can determine in principle from one single event also the mass of a particle. Therefor it was thoroughly discussed whether one could use a C-REMI to measure, in the decay process of Tritium, the neutrino momentum by performing a He^{1+} recoil ion-electron coincidence. To be sensitive to a very small neutrino mass of about one eV in this decay event the electron kinetic energy must be very close to the Q-value of the tritium decay. These events are extremely rare. Thus because of the huge number of random coincidences the required time of measurement would nearly approach the life-time of an experimenter. May be somebody will discover a way to handle such a high random rate?

6 Conclusion

The C-REMI technique can be considered as the "Bubble Chamber" or "Time Projection Chamber" in atomic and molecular physics. Using the multi-coincidence concept initially developed in nuclear and high energy particle physics, a C-REMI can image the whole momentum space in a single-event quantum process. Using ultra-cold targets in the gas phase and electro-magnetic spectrometer designs with focusing conditions an excellent sub-atomic momentum resolution and a large multi-hit coincidence efficiency are obtained. Thus visualizing the complete dynamics in a single event the dynamical entanglement in many particle systems can be explored. The C-REMI is now a standard detection system in many fields of physics and chemistry and is used by many groups around the world.

Acknowledgements Many of the scientists contributing to the C-REMI development are co-authors of this paper. But we are indebted to many other colleagues: Reinhold Schuch, Siegbert Hagmann, Amin Cassimi, Nora Berrah, Andre Staudte, Harald Bräuning, Angela Bräuning Demian, Paul Corkum, Kiyoshi Ueda, Tadashi Kambara, Yasu Yamazaki, Paul Mokler, Thomas Stöhlker, Klaus Blaum, Erhard Salzborn, Alfred Müller, Karl Ontjes Groeneveld, Hans Joachim Specht,

Bernd Sonntag, Jochen Schneider, Berthold Krässig, Timor Osipov, etc. for a long close cooperation and many theorists providing ideas for measurements like John Briggs, Burkhard Fricke, Hans Jürgen Lüdde, Jan Michael Rost, Steve Manson, etc. Additionally we want to thank BMBF (Dietrich Hartwig) at GSI, the Deutsche Forschungsgemeinschaft, the people at GSI, ALS, at Hasylab, Bessy, Grenoble, Soleil Paris, Spring8, the mechanics and electro technicians in Frankfurt, KSU, Heidelberg, GSI and Berkeley workshops for continuous support.

The early team laying the foundations of C-REMI. Painting by Jürgen Jaumann 2011

References

1. H. Schmidt-Böcking, S. Eckart, H. J. Lüdde, G. Gruber, T. Jahnke, The precision limits in a single-event quantum measurement of electron momentum and position, these proceedings
2. O. Stern, Zur Methode der Molekularstrahlen I. Z. Physik. **39**, 751–763 (1926); F. Knauer, O. Stern, Zur Methode der Molekularstrahlen II. Z. Physik. **39**, 764–779 (1926)
3. H. Schmidt-Böcking, K. Reich, Otto Stern-Physiker, Querdenker, Nobelpreisträger. Goethe-Universität Frankfurt, Herausgeber. Gründer, Gönner und Gelehrte. Societätsverlag, Reihe. ISBN 978-3-942921-23-7 (2011)
4. J. Ullrich, R. Dörner, V. Mergel, O. Jagutzki, L. Spielberger, H. Schmidt-Böcking, Cold-target recoil-ion momentum-spectroscopy: first results and future perspectives of a novel high resolution technique for the investigation of collision induced many-particle reactions. Comments Atomic Mol. Phys. **30**, 285 (1994)
5. J. Ullrich, R. Dörner, H. Schmidt-Böcking, A New "Momentum Microscope" Views Atomic Collision Dynamics, Physics News. American Institute of Physics (1996), p. 12
6. R. Moshammer, M. Unverzagt, W. Schmitt, J. Ullrich, H. Schmidt-Böcking, A 4 π recoil-ion electron momentum analyser: a high-resolution, "microscope" for the investigation of the dynamics of atomic, molecular and nuclear reactions. Nucl. Instrum. Meth. B **108**, 425 (1996)
7. J. Ullrich, R. Moshammer, R. Dörner, O. Jagutzki, V. Mergel, H. Schmidt-Böcking, L. Spielberger, Recoil ion momentum spectroscopy. J. Phys. B At. Mol. Opt. **30**, 2917 (1997); J.

Ullrich, W. Schmitt, R. Dörner, O. Jagutzki, V. Mergel, R. Moshammer, H. Schmidt-Böcking, L. Spielberger, M. Unverzagt, R.E. Olson, in Recoil Ion Momentum Spectroscopy Photonic, ed. by F. Aumayr et al. Electr. Atomic Coll., World Scient. (1997), p. 421

8. R. Dörner, V. Mergel, H. Bräuning, M. Achler, T. Weber, Kh. Khayyat, O. Jagutzki, L. Spiel-berger, J. Ullrich, R. Moshammer, Y. Azuma, M. H. Prior, C.L. Cocke, H. Schmidt-Böcking, in Recoil ion momentum spectroscopy—a "momentum microscope" to view atomic collision dynamics. Proceedings of the AIP Conference—Atomic Processes in Plasmas, ed. by E. Oks, M. Pindzola (1998), p. 443

9. R. Dörner, V. Mergel, O. Jagutzki, L. Spielberger, J. Ullrich, R. Moshammer, H. Schmidt-Böcking, Cold target recoil ion momentum spectroscopy: a "momentum microscope" to view atomic collision dynamics. Phys. Rep. **330**, 95 (2000)

10. R. Dörner, Th. Weber, M. Weckenbrock, A. Staudte, M. Hattass, R. Moshammer, J. Ullrich, H. Schmidt-Böcking, Multiple Ionization in Strong Laser Fields Advances in Atomic and Molecular Physics, vol. 48, ed. by B. Bederson, H. Walther. Academic Press (2002), p. 1

11. J. Ullrich, R. Moshammer, A. Dorn, R. Dörner, L. Ph. H. Schmidt, H. Schmidt-Böcking Recoil-ion and electron momentum spectroscopy: reaction-microscopes. Rep. Prog. Phys. **66**, 1463 (2003)

12. O. Stern, Eine direkte Messung der thermischen Molekulargeschwindigkeit. Z. Physik **2**, 49–56 (1920); O. Stern, Nachtrag zu meiner Arbeit: „Eine direkte Messung der thermischen Molekulargeschwindigkeit, Z. Physik **3**, 417–421 (1920)

13. S. Grundmann, D. Trabert, K. Fehre, N. Strenger, A. Pier, L. Kaiser, M. Kircher, M. Weller, S. Eckart, L. Ph. H. Schmidt, F. Trinter, T. Jahnke, M. S. Schöffler, R. Dörner, Zeptosecond Birth Time Delay in Molecular Photoionization. Sci. **370**, 339–341 (2020)

14. L. Dunoyer, Le Radium **8**, 142 (1911)

15. W. Gerlach, O. Stern, Der experimentelle Nachweis der Richtungsquantelung im Magnet-feld. Z. Physik, **9**, 349–352 (1922); W. Gerlach, O. Stern, Über die Richtungsquantelung im Magnetfeld. Ann. Physik, **74**, 673–699 (1924)

16. H. Schmidt-Böcking, H. Reich, K. Templeton, W. Trageser, V. Vill (Hrsg.). Otto Sterns Veröf-fentlichungen—Band 1 bis V Sterns Veröffentlichungen von 1912 bis 1916 Springer Verlag. ISBN 978-3-662-46953-8 (2016)

17. O.R. Frisch, Experimenteller Nachweis des Einsteinschen Strahlungsrückstoßes. Z. Phys. **86**, 42–48 (1933)

18. Center for History of Science, The Royal Swedish Academy of Sciences, Box 50005, SE-104 05 Stockholm, Sweden

19. S. Kelbch, H. Schmidt-Böcking, J. Ullrich, R. Schuch, E. Justiniano, H. Ingwersen, C.L. Cocke, The contributions of K-electron capture for the production of highly charged Ne recoil ions by 156 MeV bromine impact. Z. Phys. A **317**, 9 (1984)

20. J. Ullrich, C.L. Cocke, S. Kelbch, R. Mann, P. Richard, H. Schmidt-Böcking, A parasite ion source for bare-ion production on a high energy heavy-ion accelerator. J. Phys. B **17**, L 785 (1984)

21. S. Kelbch, J. Ullrich, R. Mann, P. Richard, H. Schmidt-Böcking, Cross sections for the produc-tion of highly charged Argon and Xenon recoil ions in collisions with high velocity uranium projectile. J. Phys. B **18**, 323 (1985)

22. P. Richard, J. Ullrich, S. Kelbch, H. Schmidt-Böcking, R. Mann, C.L. Cocke, The production of highly charged Ar and Xe recoil ions by fast uranium impact. Nucl. Instr. Meth. A **240**, 532 (1985)

23. H. Schmidt-Böcking, C.L. Cocke, S. Kelbch, R. Mann, P. Richard, J. Ullrich, Multiple Ioniza-tion of Argon Atoms by Fast Uranium Impact and its Possible Application as an Ion Source for Highly Ionized Rare Gas Atoms, High Energy Ion Atom Collisions, eds. by D. Berenyi, G. Hock. Akademia Kiado, Budapest (1985)

24. S. Kelbch, J. Ullrich, W. Rauch, H. Schmidt-Böcking, M. Horbatsch, R. Dreizler, S. Hagmann, R. Anholt, A.S. Schlachter, A. Müller, P. Richard, Ch. Stoller, C. L. Cocke, R. Mann, W. E. Meyerhof, J.D. Rasmussen, Multiple Ionization of Ne, Ar, Kr and I by nearly Relativistic U Ions. J. Phys. B **19**, L 47 (1986)

25. R.E. Olson, J. Ullrich, H. Schmidt-Böcking. J. Phys. B20, L809 (1987); Grandin et al. Europhys. Lett. **6**, 683 (1988)

26. C.L. Cocke, R.E. Olson, Recoil ions. Phys. Rep. **205**, 155 (1991)

27. U. Buck, M. Düker, H. Pauly, D. Rust, Proceed. of the IV Int. Symp. Molecular beams (1974) 70; H. Haberland, U. Buck, M. Tolle. Rev. Sci. Instr. **56**, 1712 (1985)

28. M. van der Poel, C.V. Nielsen, M.A. Gearba, N. Andersen. Phys. Rev. Lett. **87**, 123–201 (2001); J.W. Turkstra, R. Hoekstra, S. Knoop, D. Meyer, R. Morgenstern, R.E. Olson, Phys. Rev. Lett. **87**, 123–202 (2001); X. Flechard, H. Nguyen, E. Wells, I. Ben-Itzhak, B.D. DePaola, Phys. Rev. Lett. **87**, 123–203 (2001)

29. R. Mann, F. Folkmann, K.O. Groeneveld, strong molecular effects in heavy-ion-induced carbon and nitrogen auger transitions. Phys. Rev. Lett. 1674 (1976); R. Mann, C.L. Cocke, A.S. Schlachter, M. Prior, R. Marrus, Selective final-state population in electron capture by low-energy highly charged projectiles studied by energy-gain spectroscopy. Phys. Rev. Lett. 1329 (1982)

30. G. Gaukler, H. Schmidt-Böcking, R. Schuch, R. Schulé, H.J. Specht, I. Tserruya, A position sensitive parallel plate avalanche detector for heavy-ion X-ray coincidence measurements. Nucl. Instr. Meth. **141**, 115 (1977)

31. J.L. Wiza, Microchannel plate detectors. Nucl. Instrum. Meth. **162**, 587–601 (1979)

32. C. Martin, P. Jelinsky, M. Lampton, R.F. Malina, H.O. Anger, Rev. Sci. Instrum. **52**, 1067 (1981)

33. O. Jagutzki, V. Mergel, K. Ullmann-Pfleger, L. Spielberger, U. Meyer, R. Dörner, H. Schmidt-Böcking, Fast position and time resolved read-out of micro-channelplates with the delay-line technique for single particle and photon detection, imaging spectroscopy IV. Proceedings of the International Symposium on Optimal Science Engineering & Instr., eds. by M.R. Descour, S.S. Shen, vol. 3438. Proc SPIE (1998), pp. 322–333; K. Fehre, D. Trojanowskaja, J. Gatzke, M. Kunitski, F. Trinter, S. Zeller, L. Ph. H. Schmidt, J. Stohner, R. Berger, A. Czasch, O. Jagutzki, T. Jahnke, R. Dörner, M.S. Schöffler, Absolute ion detection efficiencies of microchannel plates and funnel microchannel plates for multi-coincidence detection. Rev. Sci. Instrum. **89**, 045112 (2018)

34. S.E. Sobottka, M.B. Williams, IEEE Trans. Nucl. Sci. **35**, 348 (1988)

35. http://roentdek.com/detectors/ M. S. Schöffler, Grundzustandskorrelationen und dynamische Prozesse untersucht in Ion-Helium-Stößen, Dissertation(2006) Universität Frankfurt

36. https://aktuelles.uni-frankfurt.de/menschen/vom-selbstgebauten-pc-zum-supercomputer/; https://www.gsi.de/work/kurier/Ausgabe/19;2018.htm?no_cache=1&cHash=4fe9358d35b5 f37e110eefdd2d86e1b1

37. K. Ullmann, V. Mergel, L. Spielberger, T. Vogt, U. Meyer, R. Dörner, O. Jagutzki, M. Unverzagt, I. Ali, J. Ullrich, W. Schmitt, R. Moshammer, C.L. Cocke, T. Kambara, Y. Awaya, H. Schmidt-Böcking, Cold target recoil ion momentum spectroscopy, in Proceedings of the 4th US-Mexican Symposium on Atomic and Molecular Physics, eds. by I. Alvarez, C. Cisneros, T.J. Morgan. World Scientific (1995), p. 269

38. https://groups.nscl.msu.edu/nscl_library/manuals/eggortec/453.pdf

39. https://www.cronologic.de/

40. J. Ullrich, H. Schmidt-Böcking, S. Kelbch, C. Kelbch, V. Dangendorf, A. Visser, D. Weisinger, Stoßparameterabhängigkeit der Vielfachionisationswahrscheinlichkeit. Annual Report Institute for Nuclear Physics, University Frankfurt, (1984), p. 20

41. J. Ullrich, H. Schmidt-Böcking, Time of flight spectrometer for the determination of micro-radian projectile scattering angles in atomic collisions. Phys. Lett. A **125**, 193 (1987)

42. R.E. Olson, J. Ullrich, R. Dörner, H. Schmidt-Böcking, Single and double ionization cross sections for angular scattering of fast protons by helium. Phys. Rev. A **40**, R2843 (1989)

43. J. Ullrich, R. Olson, R. Dörner, V. Dangendorf, S. Kelbch, H. Berg, H. Schmidt-Böcking, Influence of ionized electrons on heavy nuclei angular differential scattering cross section. J. Phys. B-At. Mol. Opt. **22**, 627 (1989)

44. R. Dörner, J. Ullrich, R.E. Olson, H. Schmidt-Böcking, Three-body interactions in proton-helium angular scattering. Phys. Rev. Lett. **63**, 147 (1989)

45. C.E. Gonzales Lepra, M. Breining, J. Burgdörfer, R. DeSerio, S.B. Elston, J.P. Gibbons, H.P. Hülskötter, L. Liljeby, R.T. Short, C.R. Vane, Nucl. Instr. Meth. B24/25 (1987), p. 316

46. J.C. Levin, R.T. Short, C.S. O., H. Cederquist, S.B. Elston, J.P. Gibbons, I.A. Sellin, H. Schmidt-Böcking, Steep dependence of recoil-ion energy on coincident projectile and target ionization in swift ion-atom collisions. Phys. Rev. **A36**, 1649 (1987)
47. I.A. Sellin, J.C. Levin, O.C.-S., H. Cederquist, S.B. Elston, R.T. Short, H. Schmidt-Böcking, Cold highly ionized ions: comparison of energies of recoil ions produced by heavy ions and by synchrotron radiation x-rays. Physica Scripta **T22**, 178–182 (1988)
48. R. Ali, V. Frohne, C.L. Cocke, M. Stöckli, S. Cheng, M.L.A. Raphaelian, Phys. Rev. Lett. **69**, 2491 (1992)
49. V. Frohne, S. Cheng, R. Ali, M. Raphaeilien, C.L. Cocke, R.E. Olson, Phys. Rev. Lett. **71**, 696 (1993)
50. W.C. Wiley, I.H. McLaren, Time-of-flight mass spectrometer with improved resolution. Rev. Sci. Instrum. **26**, 1150 (1955)
51. R. Dörner, J. Ullrich, O. Jagutzki, S. Lencinas, A. Gensmantel, H. Schmidt-Böcking, in Electronic and Atomic Collisions, ed. by W.R. MacGillivray, I.E. McCarthy, and M.C. Standage (Adam Hilger, Bristol, 1992), p. 351
52. V. Mergel, R. Dörner, J. Ullrich, O. Jagutzki, S. Lencinas, S. Nüttgens, L. Spielberger, M. Unverzagt, C.L. Cocke, R. E. Olson, M. Schulz, U. Buck, E. Zanger, W. Theisinger, M. Isser, S. Geis, H. Schmidt- Böcking, State selective scattering angle dependent capture cross sections using cold target recoil ion momentum spectroscopy (COLTRIMS), Phys. Rev. Lett. **74**, 2200 (1995); Diplomarbeit (1994), Institute f. Nucl. Physics, University Frankfurt
53. A. Cassimi, S. Duponchel, X. Flechard, P. Jardin, P. Sortais, D. Hennecart, R.E. Olson, Phys. Rev. Lett. **76**, 20 3679 (1996)
54. J.P. Grandin, D. Hennecart, X. Husson, D. Lecler, I. Lesteven-Vaisse, D. Lisfi, Europhys. Lett. **6**, 683 (1988)
55. P. Jardin, J.P. Grandin, A. Cassimi, J.P. Lemoigne, A. Gosslin, X. Husson, D. Hennecart, A. Lepontre, in 5th. Conference on Atomic Physics of Highly Charged Ions (AIP Proceed. 274) (1990), p. 291
56. P. Jardin, A. Cassimi, J.P. Grandin, H. Rothard, J.P. Lemoigne, A. Gosslin, X. Husson, D. Hennecart, A. Lepontre, Nucl. Instr. Meth. **B 107** (1996), p. 41
57. Dominique Akoury. Photodoppelionisation von molekularem Wasserstoff bei hohen Photonenenergien, Diplomarbeit (2008) https://www.atom.uni-frankfurt.de/publications/
58. M.S. Schöffler, Grundzustandskorrelationen und dynamische Prozesse untersucht in Ion-Helium-Stößen, Dissertation, Universität Frankfurt, and private communication, 2006
59. N.V. Federenko, V.V. Afrosimov, Sov. Phys.-Tech. Phys. **1**, 1872 (1956) 1956
60. L.J. Puckett, D.W. Martin, Phys. Rev. A **5**, 1432 (1976)
61. W. Steckelmacher, R. Strong, M.W. Lucas, J. Phys. **B11** (1978), 1553; W. Steckelmacher, R. Strong, M.W. Lucas, A simple atomic or molecular beam as target for ion-atom collision studies. J. Phys. D: Appl. Phys. **11**, 1553 (1978)
62. B. Fastrup, Inelastic Energy-Loss Measurements in Single Collisions, in Methods of Experimental Physics, vol. 17. (Academic Press, 1980), p. 149
63. X. Flechard, C. Harel, H. Jouin, B. Pons, L. Adoui, F. Freemont, A. Cassimi, D. Hennecart, J. Phys. B. **34**, 2759 (2001)
64. Th Weber, Kh Khayyat, R. Dörner, V.D. Rodriguez, V. Mergel, O. Jagutzki, L. Schmidt, K.A. Müller, F. Afaneh, A. Gonzales, H. Schmidt-Böcking, Abrupt rise of the longitudinal recoil ion momentum distribution for ionizing collisions. Phys. Rev. Lett. **86**, 224 (2001)
65. D. Fischer, B. Feuerstein, R.D. Dubois, R. Moshammer, J.R. Crespo Lopèz-Urrutia, I. Draganic, H. Lörch, A.N. Perumal, J. Ullrich, J. Phys. B **35**, 1369 (2002); H.K. Kim, M.S. Schöffler, S. Houamer, O. Chuluunbaatar, J. N. Titze, L.Ph.H. Schmidt, T. Jahnke, H. Schmidt-Böcking, A. Galstyan, Yu.V. Popov, R. Dörner, Electron transfer in fast proton-helium collisions. Phys. Rev. A, **85**, 022707 (2012)
66. R. Dörner, V. Mergel, R. Ali, U. Buck, C.L. Cocke, K. Froschauer, O. Jagutzki, S. Lencinas, W.E. Meyerhof, S. Nüttgens, R.E. Olson, H. Schmidt-Böcking, L. Spielberger, K. Tökesi, J. Ullrich, M. Unverzagt, W. Wu, Electron-electron interaction in projectile ionization investigated by high resolution recoil ion momentum spectroscopy. Phys. Rev. Lett. **72**, 3166 (1994)

67. W. Wu, K.L. Wong, R. Ali, C.Y. Chen, C.L. Cocke, V. Frohne, J.P. Giese, M. Raphaelian, B. Walch, R. Dörner, V. Mergel, H. Schmidt-Böcking, W.E. Meyerhof, Experimental separation of electron-electron and electron-nuclear contributions to ionisation of fast hydrogenlike ions colliding with He. Phys. Rev. Lett. **72**, 3170 (1994); W. Wu, K.L. Wong, E.C. Montenegro, R. Ali, C.Y. Chen, C.L. Cocke, R. Dörner, V. Frohne, J.P. Giese, V. Mergel, W.E. Meyerhof, M. Raphaelian, H. Schmidt-Böcking, B. Walch, Electron-electron interaction in the ionization of O^{7+} by He. Phys. Rev. A **55**, 2771 (1997)

68. R. Moshammer, J. Ullrich, M. Unverzagt, W. Schmidt, P. Jardin, R.E. Olson, R. Mann, R. Dörner, V. Mergel, U. Buck, H. Schmidt-Böcking, Low-energy electrons and their dynamical correlation with the recoil-ions for single ionization of helium by fast, heavy-ion impact. Phys. Rev. Lett. **73**, 3371 (1994)

69. R. Moshammer private communication, GSI report (1997)

70. J. Ullrich, R. Dörner, V. Mergel, O. Jagutzki, L. Spielberger, H. Schmidt-Böcking, Cold-target recoil-ion momentum-spectroscopy: first results and future perspectives of a novel high resolution technique for the investigation of collision induced many-particle reactions. Comments Atomic Mol. Phys. **30**, 285 (1994)

71. M. Unverzagt, R. Moshammer, W. Schmitt, R. E. Olson, P. Jardin, V. Mergel, H. Schmidt-Böcking, Collective behavior of electrons emitted in multiply ionizing of 5.9 MeV/m U^{65+} with Ne. Phys. Rev. Lett. **76**, 1043 (1996)

72. R. Moshammer, J. Ullrich, H. Kollmus, W. Schmitt, M. Unverzagt, O. Jagutzki, V. Mergel, H. Schmidt-Böcking, R. Mann, C. Woods, R.E. Olson, Double ionization of helium and neon for fast heavy-ion impact: correlated motion of electrons from bound in continuum states. Phys. Rev. Lett. **77**, 1242 (1996)

73. R. Moshammer, J. Ullrich, H. Kollmus, W. Schmitt, M. Unverzagt, H. Schmidt-Böcking, C.J. Wood, R.E. Olson, Complete momentum balance for single ionization of helium by fast ion impact: experiment. Phys. Rev. A **56**, 1351 (1997)

74. R. Moshammer, W. Schmitt, J. Ullrich, H. Kollmus, A. Cassimi, R. Dörner, O. Jagutzki, R. Mann, R.E. Olson, H.T. Prinz, H. Schmidt-Böcking, L. Spielberger, Ionization of helium in the attosecond equivalent light pulse of 1 GeV/Nucleon U^{92+} projectiles. Phys. Rev. Lett. **79**, 3621 (1997)

75. M. Schulz, R. Moshammer, D. Fischer, H. Kollmus, D.H. Madison, S. Jones, J. Ullrich, Three-dimensional imaging of atomic four-body processes. Nature **422**, 48–50 (2003)

76. R. Dörner, T. Vogt, V. Mergel, H. Khemliche, S. Kravis, C.L. Cocke, J. Ullrich, M. Unverzagt, L. Spielberger, M. Damrau, O. Jagutzki, I. Ali, B. Weaver, K. Ullmann, C.C. Hsu, M. Jung, E.P. Kanter, B. Sonntag, M.H. Prior, E. Rotenberg, J. Denlinger, T. Warwick, S.T. Manson, H. Schmidt-Böcking, Ratio of cross sections for double to single ionization of He by 85-400 eV photons. Phys. Rev. Lett. **76**, 2654 (1996)

77. H. Schmidt-Böcking, C.L. Cocke, R. Dörner, O. Jagutzki, T. Kambara, V. Mergel, R. Moshammer, M.H. Prior, L. Spielberger, W. Schmitt, K. Ullmann, M. Unverzagt, J. Ullrich, W. Wu, in Accelerator-based atomic physics techniques and application, ed. by S.M. Shafroth, J.C. Austin (AIP, Woodbury, New York, 1996), pp. 723–745

78. L. Spielberger, O. Jagutzki, R. Dörner, J. Ullrich, U. Meyer, V. Mergel, M. Unverzagt, M. Damrau, T. Vogt, I. Ali, Kh. Khayyat, D. Bahr, H. G. Schmidt, R. Frahm, H. Schmidt-Böcking, Separation of photoabsorption and compton scattering contributions to He single and double ionization. Phys. Rev. Lett. **74**, 4615 (1995); L. Spielberger, O. Jagutzki, B. Krässig, U. Meyer, Kh. Khayyat, V. Mergel, Th. Tschentscher, Th. Buslaps, H. Bräuning, R. Dörner, T. Vogt, M. Achler, J. Ullrich, D.S. Gemmell, H. Schmidt-Böcking, Double and single ionization of helium by 58-keV X-Rays. Phys. Rev. Lett. **76**, 4685 (1996)

79. O. Schwarzkopf, B. Krässig, J. Elminger, V. Schmidt, Phys. Rev. Lett. **70**, 3008 (1993)

80. A. Huetz, P. Laplanque, L. Andric, P. Selles, J. Mazeau, J. Phys. B: At. Mol. Opt. Phys. **27**, L13 (1994)

81. U. Becker, D. Shirley, in: VUV and Soft X-ray Photoionization (Chapter 5). (Springer, 2012)

82. R. Dörner, J. Feagin, C.L. Cocke, H. Bräuning, O. Jagutzki, M. Jung, E.P. Kanter, H. Khemliche, S. Kravis, V. Mergel, M.H. Prior, H. Schmidt-Böcking, L. Spielberger, J. Ullrich, M. Unverzagt, T. Vogt, Fully differential cross sections for double photoionization of He measured

by recoil ion momentum spectroscopy. Phys. Rev. Lett. **77**, 1024 (1996)

83. R. Dörner, H. Bräuning, O. Jagutzki, V. Mergel, M. Achler, R. Moshammer, J.M. Feagin, T. Osipov, A. Bräuning-Demian, L. Spielberger, J.H. McGuire, M.H. Prior, N. Berrah, J.D. Bozek, C.L. Cocke, H. Schmidt-Böcking, Double photoionization of spatially aligned D_2. Phys. Rev. Lett. **81**, 5776 (1998)

84. V. Mergel, M. Achler, R. Dörner, Kh. Khayyat, T. Kambara, Y. Awaya, V. Zoran, B. Nyström, L. Spielberger, J.H. McGuire, J. Feagin, J. Berakdar, Y. Azuma H. Schmidt-Böcking, Helicity dependence of the photon-induced three-body coulomb fragmentation of helium investigated by COLTRIMS. Phys. Rev. Lett. **80**, 5301 (1998)

85. R. Dörner, H. Khemliche, M.H. Prior, C.L. Cocke, J.A. Gary, R.E. Olson, V. Mergel, J. Ullrich, H. Schmidt-Böcking, Imaging of saddle point electron emission in slow p-He collisions. Phys. Rev. Lett. **77**, 4520 (1996)

86. E. Horsdal-Pedersen, C.L. Cocke, M. Stöckli, Phys. Rev. Lett. **50**, 1910 (1983)

87. H. Vogt, R. Schuch, E. Justiniano, M. Schulz, W. Schwab, Phys. Rev. Lett. **57**, 2250 (1986)

88. L.H. Thomas, Proc. R. Soc. London A **114**, 561 (1927)

89. E. Horsdal-Pedersen, B. Jensen, K.O. Nielsen, Phys. Rev. Lett. **57**, 1414 (1986)

90. R. Shakeshaft, L. Spruch, Rev.Mod.Phys. **51**, 369 (1979)

91. V. Mergel, R. Dörner, M. Achler, Kh. Khayyat, S. Lencinas, J. Euler, O. Jagutzki, S. Nüttgens, M. Unverzagt, L. Spielberger, W. Wu, R. Ali, J. Ullrich, H. Cederquist, A. Salin, C.J. Wood, R.E. Olson, Dz. Belkic, C.L. Cocke, H. Schmidt-Böcking, Intra-atomic electron-electron scattering in p-He collisions (Thomas process) investigated by cold target recoil ion momentum spectroscopy. Phys. Rev. Lett. **79**, 387 (1997); V. Mergel, Ph.D. thesis, University Frankfurt. ISBN 3-8265-2067-X (1996) unpublished

92. V. Mergel, R. Dörner, Kh Khayyat, M. Achler, T. Weber, H. Schmidt-Böcking, H.J. Lüdde, Strong correlations in the He ground state momentum wave function—observed in the fully differential momentum distributions for the four particle p + He transfer ionization process. Phys. Rev. Lett. **86**, 2257 (2001)

93. M. Schöffler, A.L. Godunov, C.T. Whelan, H.R.J. Walters, V.S. Schipakov, V. Mergel, R. Dörner, O. Jagutzki, L.Ph.H Schmidt, J. Titze, E. Weigold, H. Schmidt-Böcking, Revealing the effect of angular correlation in the ground-state He wavefunction: a coincidence study of the transfer ionization process. J. Phys. B-At. Mol. Opt. **38**, L123 (2005)

94. M.S. Schöffler, Grundzustandskorrelationen und dynamische Prozesse untersucht in Ion-Helium-Stößen, Dissertation (2006); M.S. Schöffler, O. Chuluunbaatar, Yu. V. Popov, S. Houamer, J. Titze, T. Jahnke, L. Ph. H. Schmidt, O. Jagutzki, A.G. Galstyan, A.A. Gusev, Transfer ionization and its sensitivity to the ground-state wave function. Phys. Rev. A, **87**, 032715 (2013); M.S. Schöffler, O. Chuluunbaatar, S. Houamer, A. Galstyan, J.N. Titze, L.Ph.H. Schmidt, T. Jahnke, H. Schmidt-Böcking, R. Dörner, Yu.V. Popov, A.A. Gusev, C. Dal Cappello, Two-dimensional electron-momentum distributions for transfer ionization in fast proton-helium collisions. Phys. Rev. A, **88**, 042710 (2013)

95. A.L. Godunov, C.T. Whelan, H.R.J. Walters, V.S. Schipakov, M. Schöffler, V. Mergel, R. Dörner, O. Jagutzki, LPhH Schmidt, J. Titze, H. Schmidt-Böcking, Transfer ionization process p + He → H + He^{2+} + e with the ejected electron detected in the plane perpendicular to the incident beam direction. Phys. Rev. A **71**, 052712 (2005)

96. P.B. Corkum, Phys. Rev. Lett. **71**, 1994 (1993); K.C. Kulander, J. Cooper, K.J. Schafer, Phys. Rev. A **51**, 561 (1995)

97. R. Moshammer, B. Feuerstein, W. Schmitt, A. Dorn, C. D. Schroeter, J. Ullrich, H. Rottke, C. Trump, M: Wittmann, G. Korn, K. Hoffmann, W. Sandner et al. Momentum distribution of Ne ions created by an intense utrashort laser pulse, Phys. Rev. Lett. 84, 447-450 (2000); R. Moshammer, J. Ullrich, B. Feuerstein, D. Fischer, A. Dorn, C.D. Schröter, J.R. Crespo López-Urrutia, C. Höhr, H. Rottke, C. Trump, M. Wittmann, G. Korn, W. Sandner, Rescattering of ultra-low energy electrons for single ionization of Ne in the tunneling regime. Phys. Rev. Lett. **91**, 113002 (2003)

98. Th. Weber, M. Weckenbrock, A. Staudte, L. Spielberger, O. Jagutzki, V. Mergel, F. Afaneh, G. Urbasch, M. Vollmer, H. Giessen, R. Dörner, Recoil-ion momentum distributions for single and double ionization of helium in strong laser fields. Phys. Rev. Lett. **84**, 443 (2000); Th.

Weber, H. Giessen, M. Weckenbrock, G. Urbasch, A. Staudte, L. Spielberger, O. Jagutzki, V. Mergel, M. Vollmer, R. Dörner, Correlated electron emission in multiphoton double ionization. Nature, **405**, 658 (2000)

99. R. Moshammer, J. Ullrich, B. Feuerstein, D. Fischer, A. Dorn, C.D. Schröter, J.R. Crespo López-Urrutia, C. Höhr, H. Rottke, C. Trump, M. Wittmann, G. Korn, W. Sandner, Strongly directed electron emission in non-sequential double ionization of Ne by intense Laser pulses, Journal of Physics B-Atomic, Molecular and Optical Physics 36 (2003) L113-119; M. Weckenbrock, D. Zeidler, A. Staudte, Th. Weber, M. Schöffler, M. Meckel, S. Kammer, M. Smolarski, O. Jagutzki, V.R. Bhardwaj, D.M. Rayner, D.M. Villeneuve, P.B. Corkum, R. Dörner, Fully differential rates for femtosecond multiphoton double ionization of neon. Phys. Rev. Lett. **92**, 213002 (2004)

100. J. Burgdörfer, C. Lemmel, X. Tong, Invited lecture at ICPEAC 2019. arXiv:2001.02900v1 [quant-ph] 9 Jan 2020

101. D. Zeidler, A. Staudte, A.B. Bardon, D.M. Villeneuve, R. Dörner, P.B. Corkum, Controlling attosecond double ionization dynamics via molecular alignment. Phys. Rev. Lett. **95**, 203003 (2005)

102. P. Eckle, A. Pfeiffer, C. Cirelli, A. Staudte, R. Dörner, H.G. Muller, M. Büttiker, U. Keller, Attosecond ionization and tunneling delay time measurements. Science **322**, 1525 (2008)

103. M. Kress, T. Löffler, M.D. Thomson, R. Dörner, H. Gimpel, K. Zrost, T. Ergler, R. Moshammer, U. Morgner, J. Ullrich, H.G. Roskos, Determination of the carrier-envelope phase of few-cycle laser pulses with terahertz-emission spectroscopy. Nat. Phys. **2**, 327 (2006)

104. M. Dürr, A. Dorn, J. Ullrich, S.P. Cao, A. Czasch, A.S. Kheifets, J.R. Götz, J.S. Briggs, (e,3e) on helium at low impact energy: the strongly correlated three-electron continuum. Phys. Rev. Lett. **98**, 193201 (2007)

105. R. Moshammer, Y.H. Jiang, L. Foucar, A. Rudenko, Th Ergler, C.D. Schröter, S. Lüdemann, K. Zrost, D. Fischer, J. Titze, T. Jahnke, M. Schöffler, T. Weber, R. Dörner, T.J.M. Zouros, A. Dorn, T. Ferger, K.U. Kühnel, S. Düsterer, R. Treusch, P. Radcliffe, E. Plönjes, J. Ullrich, Few-photon multiple ionization of Ne and Ar by strong free-electron-laser pulses. Phys. Rev. Lett. **98**, 203001 (2007)

106. A. Staudte, C. Ruiz, M. Schöffler, S. Schössler, D. Zeidler, Th Weber, M. Meckel, D.M. Villeneuve, P.B. Corkum, A. Becker, R. Dörner, Binary and recoil collisions in strong field double ionization of helium. Phys. Rev. Lett. **99**, 263002 (2007)

107. X.-J. Liu, H. Fukuzawa, T. Teranishi, A. De Fanis, M. Takahashi, H. Yoshida, A. Cassimi, A. Czasch, L. Schmidt, R. Dörner, K. Wang, B. Zimmermann, V. McKoy, I. Koyano, N. Saito, K. Ueda, Breakdown of the two-step model in K-shell photoemission and subsequent decay probed by the molecular-frame photoelectron angular distributions of CO_2. Phys. Rev. Lett. **101**, 083001 (2008)

108. A. Rudenko, L. Foucar, M. Kurka, Th Ergler, K.U. Kühnel, Y.H. Jiang, A. Voitkiv, B. Najjari, A. Kheifets, S. Lüdemann, T. Havermeier, M. Smolarski, S. Schössler, K. Cole, M. Schöffler, R. Dörner, S. Düsterer, W. Li, B. Keitel, R. Treusch, M. Gensch, C.D. Schröter, R. Moshammer, J. Ullrich, Recoil-ion momentum distributions for two-photon double ionization of He and Ne by 44 eV free-electron laser radiation. Phys. Rev. Lett. **101**, 073003 (2008)

109. M. Meckel, D. Comtois, D. Zeidler, A. Staudte, D. Pavicic, H.C. Bandulet, H. Pépin, J.C. Kieffer, R. Dörner, D.M. Villeneuve, P.B. Corkum, Laser-induced electron tunneling and diffraction. Science **320**, 1478 (2008)

110. H. Sann, T. Havermeier, C. Müller, H.-K. Kim, F. Trinter, M. Waitz, J. Voigtsberger, F. Sturm, T. Bauer, R. Wallauer, D. Schneider, M. Weller, C. Goihl, J. Tross, K. Cole, J. Wu, M.S. Schöffler, H. Schmidt-Böcking, T. Jahnke, M. Simon, R. Dörner, Imaging the temporal evolution of molecular orbitals during ultrafast dissociation. Phys. Rev. Lett. **117**, 243002 (2016)

111. H. Kang, K. Henrichs, M. Kunitski, Y. Wang, X. Hao, K. Fehre, A. Czasch, S. Eckart, L. Ph. H Schmidt, M. Schöffler, T. Jahnke, X. Liu, R. Dörner, Timing recollision in nonsequential double ionization by intense elliptically polarized laser pulses. Phys. Rev. Lett. **120**, 223204 (2018)

112. Y.H. Jiang, A. Rudenko, M. Kurka, K.U. Kühnel, Th Ergler, L. Foucar, M. Schöffler, S. Schössler, T. Havermeier, M. Smolarski, K. Cole, R. Dörner, S. Düsterer, R. Treusch, M. Gensch, C.D. Schröter, R. Moshammer, J. Ullrich, Few-photon multiple ionization of N2 by

extreme ultraviolet free-electron laser radiation. Phys. Rev. Lett. **102**, 123002 (2009)

113. E. Shigemasa, J. Electron Spectrosc. Relat. Phenom. **88–91**, 9 (1998)

114. F. Heiser, O. Gessner, J. Viefhaus, K. Wieliczek, R. Hentges, U. Becker, Phys. Rev. Lett. **79**, 2435 (1997)

115. https://www.researchgate.net/scientific-contributions/31908446_P_M_Guyon ZEKE

116. A. Landers, Th Weber, I. Ali, A. Cassimi, M. Hattass, O. Jagutzki, A. Nauert, T. Osipov, A. Staudte, M.H. Prior, H. Schmidt-Böcking, C.L. Cocke, R. Dörner, Photoelectron diffraction mapping: molecules illuminated from within. Phys. Rev. Lett. **87**, 013002 (2001)

117. A. Lafosse, M. Lebech, J.C. Brenot, P.M. Guyon, O. Jagutzki, L. Spielberger, M. Vervloet, J.C. Houver, D. Dowek, Vector correlations in dissociative photoionization of diatomic molecules in the VUV range: strong anisotropies in electron emission from spatially oriented NO molecule. Phys. Rev. Lett. **84**, 5987 (2000)

118. T. Jahnke, Th. Weber, A.L. Landers, A. Knapp, S. Schössler, J. Nickles, S. Kammer, O. Jagutzki, L. Schmidt, A. Czasch, T. Osipov, E. Arenholz, A.T. Young, R. Diez Muino, D. Rolles, F.J. Garcia de Abajo, C.S. Fadley, M.A. Van Hove, S.K. Semenov, N.A. Cherepkov, J. Rösch, M.H. Prior, H. Schmidt-Böcking, C.L. Cocke, R. Dörner, Circular dichroism in K-shell ionization from fixed-in-space CO and N_2 molecules. Phys. Rev. Lett. **88**, 073002 (2002)

119. K. Ueda, A. De Fanis, N. Saito, M. Machida, K. Kubozuka, H. Chiba, Y. Muramatu, Y. Sato, A. Czasch, O. Jaguzki, R. Dörner, A. Cassimi, M. Kitajima, T. Furuta, H. Tanaka, S.L. Sorensen, K. Okada, S. Tanimoto, K. Ikejiri, Y. Tamenori, H. Ohashi, I. Koyano, Nuclear motion and symmetry breaking of the B 1S-exited BF3 molecule. Chem. Phys. **289**, 135 (2003)

120. T. Jahnke, L. Foucar, J. Titze, R. Wallauer, T. Osipov, E. P. Benis, A. Alnaser, O. Jagutzki, W. Arnold, S. K. Semenov, N. A. Cherepkov, L. Ph. H. Schmidt, A. Czasch, A. Staudte, M. Schöffler, C. L. Cocke, M.H. Prior, H. Schmidt-Böcking, R. Dörner, Vibrationally resolved K-shell photoionization of CO with circularly polarized light. Phys. Rev. Lett. **93**, 083002 (2004)

121. F. Trinter, L.Ph.H. Schmidt, T. Jahnke, M.S. Schöffler, O. Jagutzki, A. Czasch, J. Lower, T.A. Isaev, R. Berger, A.L. Landers, Th. Weber, R. Dörner, H. Schmidt-Böcking, Multifragment vector correlation imaging. A search for hidden dynamical symmetries in many-particle molecular fragmentation processes. Mol. Phys. **110**, 1863 (2012)

122. Th. Weber, M. Weckenbrock, M. Balser, L. Schmidt, O. Jagutzki, W. Arnold, O. Hohn, M. Schöffler, E. Arenholz, T. Young, T. Osipov, L. Foucar, A. De Fanis, R. Diez Muino, H. Schmidt-Böcking, C.L. Cocke, M.H. Prior, R. Dörner, Auger electron emission from fixed-in-space CO. Phys. Rev. Lett. **90**, 153003 (2003); T. Osipov, C.L. Cocke, M.H. Prior, A. Landers, T. Weber, O. Jagutzki, L. Ph. H. Schmidt, H. Schmidt-Böcking, R. Dörner, Photoelectron-photo-ion momentum spectroscopy as a clock for chemical rearrangements: isomerization of the dication of acetylene to the vinylidene configuration. Phys. Rev. Lett. **90**, 233002 (2003); T. Weber, A.O. Czasch, O. Jagutzki, A.K. Müller, V. Mergel, A. Kheifets, E. Rotenberg, G. Meigs, M.H. Prior, S. Daveau, A. Landers, C.L. Cocke, T. Osipov, R. Diez Muino, H. Schmidt-Böcking, R. Dörner, Complete photo-fragmentation of the deuterium molecule. Nature, **431**, 437 (2004); Th. Weber, A. Czasch, O. Jagutzki, A. Müller, V. Mergel, A. Kheifets, J. Feagin, E. Rotenberg, G. Meigs, M.H. Prior, S. Daveau, A.L. Landers, C.L. Cocke, T. Osipov, H. Schmidt-Böcking, R. Dörner, Fully differential cross sections for photo-double-ionization of D2. Phys. Rev. Lett. **92**, 163001 (2004); F. Martín, J. Fernández, T. Havermeier, L. Foucar, Th. Weber, K. Kreidi, M. Schöffler, L. Schmidt, T. Jahnke, O. Jagutzki, A. Czasch, E.P. Benis, T. Osipov, A.L. Landers, A. Belkacem, M.H. Prior, H. Schmidt-Böcking, C.L. Cocke, R. Dörner, Single photon induced symmetry breaking of H2 dissociation. Science, **315**, 629 (2007); J.B. Williams, C.S. Trevisan, M.S. Schöffler, T. Jahnke, I. Bocharova, H. Kim, B. Ulrich, R. Wallauer, F. Sturm, T.N. Rescigno, A. Belkacem, R. Dörner, Th. Weber, C.W. McCurdy, A.L. Landers, Imaging polyatomic molecules in three dimensions using molecular frame photoelectron angular distributions. Phys. Rev. Lett. **108**, 233002 (2012)

123. L.S. Cederbaum, J. Zobeley, F. Tarantelli, Phys. Rev. Lett. **79**, 4778 (1997)

124. S. Marburger, O. Kugeler, U. Hergenhahn, T. Möller, Phys. Rev. Lett. **90**, 203401 (2003)

125. T. Jahnke, A. Czasch, M.S. Schöffler, S. Schössler, A. Knapp, M. Käsz, J. Titze, C. Wimmer,

K. Kreidi, R.E. Grisenti, A. Staudte, O. Jagutzki, U. Hergenhahn, H. Schmidt-Böcking, R. Dörner, Experimental observation of interatomic coulombic decay in neon dimers. Phys. Rev. Lett. **93**, 163401 (2004)

126. T. Jahnke, A. Czasch, M. Schöffler, S. Schössler, M. Käsz, J. Titze, K. Kreidi, R. E. Grisenti, A. Staudte, O. Jagutzki, L. Ph. H. Schmidt, Th Weber, H. Schmidt-Böcking, K. Ueda, R. Dörner, Experimental separation of virtual photon exchange and electron transfer in interatomic coulombic decay of neon dimers. Phys. Rev. Lett. **99**, 153401 (2007)

127. K. Kreidi, Ph. V Demekhin, T. Jahnke, Th Weber, T. Havermeier, X.-J. Liu, Y. Morisita, S. Schössler, L. Ph. H. Schmidt, M. Schöffler, M. Odenweller, N. Neumann, L. Foucar, J. Titze, B. Ulrich, F. Sturm, C. Stuck, R. Wallauer, S. Voss, I. Lauter, H. K. Kim, M. Rudloff, H. Fukuzawa, G. Prümper, N. Saito, K. Ueda, A. Czasch, O. Jagutzki, H. Schmidt-Böcking, S. Scheit, L. S. Cederbaum, R. Dörner, Photo- and auger-electron recoil induced dynamics of interatomic coulombic decay. Phys. Rev. Lett. **103**, 033001 (2009)

128. N. Sisourat, H. Sann, N.V. Kryzhevoi, P. Kolorenč, T. Havermeier, F. Sturm, T. Jahnke, H.K. Kim, R. Dörner, L.S. Cederbaum, Interatomic electronic decay driven by nuclear motion. Phys. Rev. Lett. **105**, 173401 (2010)

129. T. Havermeier, T. Jahnke, K. Kreidi, R. Wallauer, S. Voss, M. Schöffler, S. Schössler, L. Foucar, N. Neumann, J. Titze, H. Sann, M. Kühnel, J. Voigtsberger, J.H. Morilla, W. Schöllkopf, H. Schmidt-Böcking, R.E. Grisenti, R. Dörner, Interatomic coulombic decay following photoionization of the helium dimer: observation of vibrational structure. Phys. Rev. Lett. **104**, 133401 (2010)

130. J. Titze, M. S. Schöffler, H.-K. Kim, F. Trinter, M. Waitz, J. Voigtsberger, N. Neumann, B. Ulrich, K. Kreidi, R. Wallauer, M. Odenweller, T. Havermeier, S. Schössler, M. Meckel, L. Foucar, T. Jahnke, A. Czasch, L. Ph. H. Schmidt, O. Jagutzki, R. E. Grisenti, H. Schmidt-Böcking, H. J. Lüdde, R. Dörner, Ionization dynamics of helium dimers in fast collisions with He^{2+}. Phys. Rev. Lett. **106**, 033201 (2011)

131. F. Trinter, J.B. Williams, M. Weller, M. Waitz, M. Pitzer, J. Voigtsberger, C. Schober, G. Kastirke, C. Müller, C. Goihl, P. Burzynski, F. Wiegandt, R. Wallauer, A. Kalinin, LPhH Schmidt, M.S. Schöffler, Y.-C. Chiang, K. Gokhberg, T. Jahnke, R. Dörner, Vibrationally resolved decay width of interatomic coulombic decay in HeNe. Phys. Rev. Lett. **111**, 233004 (2013)

132. T. Jahnke, U. Hergenhahn, B. Winter, R. Dörner, U. Frühling, Ph. V. Demekhin, K. Gokhberg, L. S. Cederbaum, A. Ehresmann, A. Knie, and A. Dreuw, Interatomic and Intermolecular Coulombic Decay, Chem. Rev. 120, 11295–11369 (2020)

133. B. Najjari, A.B. Voitkiv, C. Müller, Two-center resonant photoionization. Phys. Rev. Lett. **105**, 153002 (2010)

134. M.S. Schöffler, J. Titze, N. Petridis, T. Jahnke, K. Cole, LPhH Schmidt, A. Czasch, D. Akoury, O. Jagutzki, J.B. Williams, N.A. Cherepkov, S.K. Semenov, C.W. McCurdy, T.N. Rescigno, C.L. Cocke, T. Osipov, S. Lee, M.H. Prior, A. Belkacem, A.L. Landers, H. Schmidt-Böcking, Th Weber, R. Dörner, Ultrafast probing of core hole localization in N_2. Science **320**, 920 (2008)

135. P.S. Bagus, H.F. Schäfer, J. Chem. Phys. **56**, 224 (1972); L.C. Snyder, J. Chem. Phys. **55**, 95 (1971); J.F. Stanton, J. Gauss, R.J. Bartlett, J. Chem. Phys. **97**, 5554 (1992); J.A. Kintop, W.V.M. Machando, L.G. Ferreira, Phys. Rev. A **43**, 3348 (1991); D. Dill, S. Wallace, Phys. Rev. Lett. **41**, 1230 (1978); L.S. Cederbaum, W. Domcke, J. Chem. Phys. **66**, 5084 (1977)

136. V. Efimov, Phys. Lett. B **33**, 563–564 (1970)

137. R. E. Grisenti, W. Schöllkopf, J. P. Toennies, J. R. Manson, T. A. Savas, and Henry I. Smith, He-atom diffraction from nanostructure transmission gratings: The role of imperfections, Phys. Rev. A 61, 033608 (2000); W. Schöllkopf, J. P. Toennies, Nondestructive Mass Selection of Small van der Waals Clusters, Science **266**, 1345–1348 (1994)

138. S. Zeller, M. Kunitski, J. Voigtsberger, A. Kalinin, A. Schottelius, C. Schober, M. Waitz, H. Sann, A. Hartung, T. Bauer, M. Pitzer, F. Trinter, Ch. Goihl, Ch. Janke, Martin Richter, G. Kastirke, M. Weller, A. Czasch, M. Kitzler, M. Braune, R. E. Grisenti, Wieland Schöllkopf, L. Ph. H. Schmidt, M. Schöffler, J. B. Williams, T. Jahnke, R. Dörner, Imaging the He2 quantum halo state using a free electron laser, P. Natl. Acad. Sci. USA **113**, 14651 (2016)

139. M. Kunitski, S. Zeller, J. Voigtsberger, A. Kalinin, L. Ph. H. Schmidt, M. Schöffler, A. Czasch, W. Schöllkopf, R. E. Grisenti, T. Jahnke, D. Blume, R. Dörner, Observation of the Efimov state of the helium trimer. Science **348**, 551 (2015)

140. W. Cencek et al., J. Chem. Phys. **136**, 224303 (2012); D. Blume, C.H. Greene. B.D. Esry, J. Chem. **113**, 2145–2158 (2000); D. Blume, C.H. Greene. B.D. Esry, J. Chem. **141**, 069901 (E) (2014)

141. [https://de.wikipedia.org/wiki/Contergan-Skandal]

142. M. Pitzer, M. Kunitski, A.S. Johnson, T. Jahnke, H. Sann, F. Sturm, L.Ph.H. Schmidt, H. Schmidt-Böcking, R. Dörner, J. Stohner, J. Kiedrowski, M. Reggelin, S. Marquardt, A. Schießer, R. Berger, M.S. Schöffler, Direct determination of absolute molecular stereochemistry in gas phase by Coulomb explosion imaging. Science **341**, 1096 (2013); M. Pitzer, G. Kastirke, P. Burzynski, M. Weller, D. Metz, J. Neff, M. Waitz, F. Trinter, L.Ph.H Schmidt, J.B. Williams, T. Jahnke, H. Schmidt-Böcking, R. Berger, R. Dörner, M. Schöffler, Stereochemical configuration and selective excitation of the chiral molecule halothane. J. Phys. B-At. Mol. Opt. **49**, 234001 (2016); M. Pitzer, G. Kastirke, M. Kunitski, T. Jahnke, T. Bauer, C. Goihl, F. Trinter, C. Schober, K. Henrichs, J. Becht, S. Zeller, H. Gassert, M. Waitz, A. Kuhlins, H. Sann, F. Sturm, F. Wiegandt, R. Wallauer, L.Ph.H. Schmidt, A.S. Johnson, M. Mazenauer, B. Spenger, S. Marquardt, S. Marquardt, H. Schmidt-Böcking, J. Stohner, R. Dörner, M. Schöffler, R. Berger, Absolute configuration from different multifragmentation pathways in light-induced Coulomb explosion imaging. Chem. Phys. Chem. **17**, 2465 (2016); M. Pitzer, How to determine the handedness of single molecules using Coulomb explosion imaging. J. Phys. B-At. Mol. Opt. **50**, 153001 (2017); M. Pitzer, K. Fehre, M. Kunitski, T. Jahnke, L. Ph. H. Schmidt, H. Schmidt-Böcking, R. Dörner, M. Schöffler, Coulomb explosion imaging as a tool to distinguish between stereoisomers. JOVE - J. Vis Exp **126**, e56062 (2017); M. Pitzer, R. Berger, J. Stohner, R. Dörner, M. Schöffler, Investigating absolute stereochemical configuration with Coulomb explosion imaging, Chimia, **72**, 384 (2018). https://doi.org/10.2533/chimia.2018.384; K. Fehre, S. Eckart, M. Kunitski, M. Pitzer, S. Zeller, C. Janke, D. Trabert, J. Rist, M. Weller, A. Hartung, L.Ph.H. Schmidt, T. Jahnke, R. Berger, R. Dörner, M. S. Schöffler, Enantioselective fragmentation of an achiral molecule in a strong laser field. Sci. Adv., **5**(3) (2019); K. Fehre, S. Eckart, M. Kunitski, M. Pitzer, S. Zeller, C. Janke, D. Trabert, J. Rist, M. Weller, A. Hartung, L.Ph.H. Schmidt, T. Jahnke, R. Berger, R. Dörner, M.S. Schöffler. Enantioselective fragmentation of an achiral molecule in a strong laser field. Sci. Adv. **5**(3) (2019)

143. Z. Vager, R. Naaman, E.P. Kanter, Science **244**, 426–431 (1989)

144. L. Ph. H. Schmidt, T. Jahnke, A. Czasch, M. Schöffler, H. Schmidt-Böcking, R. Dörner, Spatial imaging of the H_2 vibrational wave function at the quantum limit. Phys. Rev. Lett. **108**, 073202 (2012)

145. P. Debye, Göttinger Nachrichten 1916 and A. Sommerfeld, Physikalische Zeitschrift **17**, 491–507 (1916)

146. L.Ph.H. Schmidt, C. Goihl, D. Metz, H. Schmidt-Böcking, R. Dörner, SYu. Ovchinnikov, J.H. Macek, D.R. Schultz, Vortices associated with the wave function of a single electron emitted in slow ion-atom collisions. Phys. Rev. Lett. **112**, 083201 (2014)

147. F. Trinter, J.B. Williams, M. Weller, M. Waitz, M. Pitzer, J. Voigtsberger, C. Schober, G. Kastirke, C. Müller, C. Goihl, P. Burzynski, F. Wiegandt, T. Bauer, R. Wallauer, H. Sann, A. Kalinin, LPhH Schmidt, M. Schöffler, N. Sisourat, T. Jahnke, Evolution of interatomic coulombic decay in the time domain. Phys. Rev. Lett. **111**, 093401 (2013)

148. A. Niehaus. J. Phys. B: Atom. Molec. Phys. **10**, 1977

149. P. Morin, I. Nenner, Phys. Rev. Lett. **56**, 1913 (1986)

150. J. Ullrich, R. Dörner, V. Mergel, O. Jagutzki, L. Spielberger, H. Schmidt-Böcking, Cold-target recoil-ion momentum-spectroscopy: first results and future perspectives of a novel high resolution technique for the investigation of collision induced many-particle reactions. Comments Atomic and Mol. Phys. **30**, 285 (1994)

5

Liquid Micro Jet Studies of the Vacuum Surface of Water and of Chemical Solutions by Molecular Beams and by Soft X-Ray Photoelectron Spectroscopy

Manfred Faubel

Liquid water, with a vapor pressure of 6.1 mbar at freezing point, is rapidly evaporating in high vacuum, rapidly cooling off by the evaporational cooling, and is freezing to ice almost instantly.

Historically, this was noted already in the very earliest experiments with vacuum by Otto Guericke, the inventor of a first vacuum pump, and by Robert Boyle. In a contemporary, serendipitous report, published in 1664 in Nurnberg, G. Scotus in his "Technica curiosa" [1] is listing a number of noteworthy, miraculous and curious observations within the newly created vacuum space inside a bell jar, like: light is penetrating vacuum, the sound of a ringing bell can not penetrate to the outside, a candle flame is extinguishing, but, gun powder can be ignited inside a vacuum; small animals such as a mouse are dying quickly in a evacuated bell jar vacuum while certain insects can survive for a while. And eventually, as an item in his volume 2, Chap. 15, "Experimentum XXXVIII. Aqua intra evacuatum Recipientum congelatur", (experiment no. 38. Water inside the evacuated recipient is freezing to ice). With scholar diligence G. Scotus has annotated that with the "new" mechanically improved vacuum pump of Robert Boyle the water is freezing in vacuum while it was not observed to freeze in the equipment provided by Otto Guericke to the laboratory of G. Scotus. From our modern view we can conclude that Boyle's pump did reach a vacuum better than 6.1 mbar, the vapor pressure of ice at freezing, while Guericke's original pump, known to have been operated with water lubricated seals and not yet with fat or oil, did stop pumping slightly above the freezing point vapor pressure of water.

Thus, for centuries to come—including the first 80 years of modern age "Otto Stern type experiments"—liquid aqueous solutions were not considered as suitable

M. Faubel (✉)
Max-Planck-Institut für Dynamik und Selbstorganisation, Göttingen, Germany
e-mail: manfred.faubel@ds.mpg.de; Manfred.Faubel@mpibpc.mpg.de

for research employing vacuum or any molecular beams technology or ultrahigh vacuum surface diagnostics.

1 The Free Vacuum Surface of Water Microjets

With this unfavorable veredictum for water in vacuum in mind, for advancing experiments successfully it is helpful to have a very close look at the dynamical processes on the surface of liquid water [2–4]. Here three principal problems dominate: (1) due to the high vapor pressure and the rapid evaporation a liquid water surface is always overcast by a dense gas cloud associated with the steady vapor stream into the adjacent vacuum space. At the minimum pressure of 6.1 mbar for liquid water, just above freezing, the molecular mean free path in this vapor is only $l = 10$ μm. At a first glance this is preventing molecular beam type experiments at the surface, unless the geometrical extension of a water surface experiment could be shrunk to total dimensions smaller than 10 μm. Furthermore, (2) the un-obstructed unilateral free flow of molecules leaving the liquid surface with a Maxwellian velocity distribution at an average velocity of water molecules of approximately 1000 m/s is resulting in a massive gas flux of 600 mbar L/s/cm^2 for each square-centimeter of liquid surface, requiring extremely large pumping capacities in order to sustain a moderately decent vacuum. This vapor flow density, equivalent to 0.027 Mol/(s cm^2), that is 0.48 g s^{-1} cm^{-2} H$_2$O, corresponds on the liquid side of the separating surface to a liquid water surface ablation rate of 4.8 mm/s. (3) Considering the water heat of vaporization of 40 kJ/mol the surface cooling rate by evaporation is of the order of 1 kW/cm^2 for liquid water in free vacuum, inducing a very rapid freezing to ice of the vacuum exposed liquid surface.

In a practicable, efficient way these three challenges can be resolved with the use of a fast flowing very thin cylindrical liquid jet in high vacuum, a method developed since the 1980's (Faubel, Schlemmer, Toennies) and, gratefully, first published in a Zeitschrift für Physik "Otto Stern Centenial Issue", 1988 [2].

For a thin liquid jet of water with diameter D smaller than the vapor mean free path $l_{H_2O} = 10$ μm: (1) Knudsen conditions, $D/l_{H_2O} < 0.5$, are met for collision-free "high vacuum" gas flow; (2) the total liquid surface is small and, thus, the vapor flow into the vacuum system is restricted to values lower than 0.5 mbar L/s, while (3) the rapid outflow of liquid from the thermostat controlled nozzle is continuously replacing the evaporation-cooled water surface before it can freeze to ice. One further benefit of the microjet technique is in providing an ultra-clean liquid surface due to the rapid replacement of the vacuum exposed surface section. A 1 mm long microjet surface section, flowing with a speed of 100 m/s, is replaced within 10 μsec. Usually, in surface science, for the handling of solid state surfaces, technically demanding ultra-high vacuum conditions with pressures lower than $p_{Vac} = 10^{-10}$ mbar are required to prevent surface coverage by impacting, and sticking, gas molecules during a typical experimental observation session of $t_{expos} \geq 1000$ s, according to Langmuir's surface coverage relation of: $p_{Vac} \cdot t_{expos} = 10^{-6}$ mbar s for coverage with 1 monolayer. On fast microjets, however, with short lived, continuously replacing surface the exposure time scales are $t_{expos}' = 10$ µs, only. Therefore, a background gas pressure of p_{Vac}' $= 10^{-3}$ mbar, with $p_{Vac}' \cdot t_{expos}' = 10^{-8}$ mbar s implying 1% surface coverage, is already the equivalent to perfectly clean, "ultrahigh vacuum" conditions from a point of view of surface coverage times.

In Fig. 1 the vacuum microjet setup scheme is depicted, showing a high-speed photography of a real jet, at 10 ns exposure time, and a sketched-in stream-tube along a vapor expansion line. A representative ensemble of gas molecules moving outward on a streamline progressively is increasing its available volume (blunt cone shape enclosures), indicated here for two different time instances, in order to illustrate the decrease in local gas density $n(r) \sim n_0 (r_0/r)^2$ in proportion to the quadratic radial distance r from the microjet. This results in a rapid increase in the local molecular mean free path $l = 1/(n(r) \cdot \sigma_{collision})$ in proportion to r^2. At the liquid surface position, at $r_0 = D/2$ given by the microjet diameter, the equilibrium vapor pressure density n_0 is assumed. Relevant for describing vacuum conditions (M. Knudsen 1905) is the number of molecular collisions on their path through a vacuum space: $dN = dr/l_{H_2O}(r)$ which is determined by the ratio between the thickness dr of a gas layer in relation to the mean free path. Integration yields the total number of collisions on the gas path taken from the liquid surface at r_0 to infinity and shows the condition of one molecular collision encounter on this way out is $l_{H_2O}(r_0)/D > 1$, the well-known Knudsen condition for a free molecular beam source [2, 4] This simple model also explains, that for near surface conditions $l_{H_2O} \ll D$, i.e. at larger jet diameters or at higher temperatures of the liquid water and correspondingly higher jet vapor pressure, many molecular collisions occur in the outward streaming vapor and cause the onset of a supersonic jet hydrodynamic expansion into the vacuum.

The liquid jet photography shown in Fig. 1 was taken with a comparably large nominal nozzle aperture diameter $D = 18$ µm for reasons of optical resolution in this near diffraction limit of optical photography. The optically smooth, cylindrical jet propagates for a distance of 2 mm downstream from the nozzle exit before it starts to build up spontaneous surface oscillations and decays rapidly into a stream of fine droplets. The evaporation cooling of the jet filament amounts to ≈ 1 °C in 10 µs [2, 3]. At typical jet speeds between 30 and 100 m/s the jet surface temperature changes by only a few degrees centigrade on the first two millimeters of the smooth jet section to be employed in free vacuum experiments.

2　Liquid Jet Flow and Decay Into Droplets

In clearer detail liquid microjet flow contours are redrawn in Fig. 2 [3]. After leaving the nozzle channel the liquid jet free surface diameter contracts by 10–20%, depending somewhat on the nozzle shape, on the viscosity and on the surface tension of the liquid. The jet liquid flow stays strictly laminar up to very high flow velocities well above $v_{jet} \geq 100$ m/s. As a consequence of the small geometrical size of the micro-jet the Reynolds numbers for flow of low-viscosity, water-like,

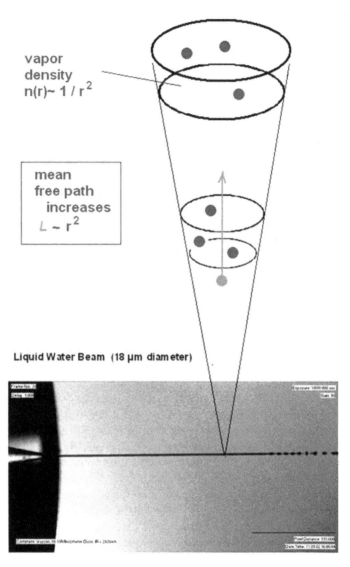

Fig. 1 A liquid microjet of water in vacuum. The photo insert shows a liquid microjet of 18 μm diameter, taken at 10 ns exposure time. The jet emerges from a quartz nozzle, and extends for several millimeters with a smooth cylindrical envelope, before decaying into a stream of droplets. Water molecules evaporating from the liquid surface rapidly are reducing in gas density with increasing distance from the liquid filament, as is indicated by the expansion stream tube cone with two control volume disks at different distances r. With decreasing densities the molecular mean free path increases. Fewer than one gas collisions occur in the emerging vapor beam when the jet diameter D is made smaller than the vapor mean free path in equilibrium, $l_{H_2O} \approx 10$ μm at T = 273 °C. This was Otto Stern's 'Knudsen condition' for operating molecular beam source ovens. By the intense vacuum evaporation the liquid jet cooling rate is ~1 °C in 10 μs, equivalent to 3 °C temperature drop on 1 mm length of flowing surface, for jets streaming with a velocity of 30 m/s

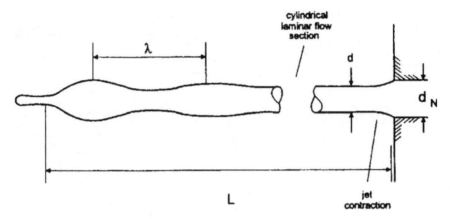

Fig. 2 Liquid jet flow and decay into droplets (figure reproduced from Ref. [3], Fig. 3)

liquids in microcapillaries are remaining well below the critical Re-values > 1500, the limit for the onset of turbulent flow [3]. This allows the production of very fast jets with smooth surfaces and long intact cylindrical sections, with moderate jet surface cooling by the powerful vacuum evaporation. Driven by surface tension (and delayed by higher liquid viscosity) the free jet filament in some distance from the nozzle begins to form contraction ripples, spontaneously, with a wavelength $\lambda \approx$ 6–8 D_{jet} and then decays rapidly into a stream of approximately uniform, equidistant droplets [3]. This decay is known as the Rayleigh spontaneous decay mode for free boundary liquid filaments. Lord Rayleigh's theory shows that the decay time to form droplets for a given liquid depends on the liquids' physical properties and on the jet diameter, only (similar to the droplet dripping time from a pipette mouth). Therefore, the contiguous length L of the smooth cylindrical section can be extended or shrunk at will, just by changing the flow speed of the jet.

A set of experimentally determined jet length values L as a function of the velocity of a liquid water jet is plotted in Fig. 3 for three different jet diameters, confirming the just discussed linear relationship between jet length L and jet velocity v_{jet} due to constant jet decay time [3]. In these measurements the nozzle was illuminated from the rear side by a laser. The light beam entered the jet like a light conducting fiber; and at the breakup point at the "end" of the liquid filament the red laser light was dispersed in all directions, creating a visible red spot which could be observed easily by a remote telescope as well for a vacuum jet as in atmosphere. The jet velocity, by Bernoulli's law, is related to the square root of the nozzle pressure $v_{jet} = Sqrt$ $(2\ p_N/\rho)$ for a low viscosity liquid of density ρ. In the diagram the jet operation pressure is shown also, in the upper ordinate scale of the plot. For the smallest jets with approximately 6 µm actual jet diameter, emerging from a nozzle with $d_N = 10$ µm aperture, the jet length increases linearly up to a maximum length $L = 4$ mm, at an approximate jet velocity near 150 m/s. At higher velocity the jet appears to decay into a diffuse, turbulent spray and the contiguous jet length begins to shorten with further increase in nozzle pressure, indicating the onset of turbulent flow and of

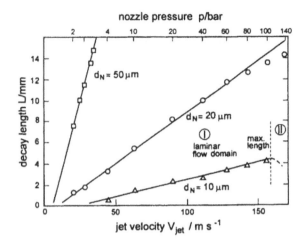

Fig. 3 Jet decay length for different jet velocities and different diameters (liquid water) (figure reproduced from Ref. [3], Fig. 4)

disruption instabilities by shearing forces. We can thus distinguish a laminar, smooth jet flow domain (I) for low velocities, which is well suited for surface experiments, from an unstable flow region (II) appearing above a certain critical speed. For the larger nozzle diameters $d_N = 20$ μm and $d_N = 50$ μm substantially longer, intact, cylindrical jet sections are obtained at much lower jet speed, as is to be expected from Rayleigh theory and its extensions, as discussed elsewhere [3, 4].

For the design of actual experiments in vacuum, microjet diameter and microjet operation conditions can be optimized and interpolated from these rough survey data in Fig. 3. From extrapolation of the jet length data shown here, it may be noted, also, that for very narrow, low viscosity water jets in the range of 1 μm diameter and smaller, the maximum decay length is decreasing dramatically to less than 0.1 mm and shorter, limiting their value for surface probing experiments with standard size microprobe devices.

3 Nascent Velocity Distribution of Evaporating Molecules

The first experiment for exploring the water microjet concept in vacuum were measurements of the velocity distribution of water molecules evaporating from the liquid microjet surface, in a set-up illustrated by Fig. 4a. [2, 3]. The liquid jet here is passing in front of a skimmer collimator entrance of a molecular beam time-of-flight spectrometer. Through a 5 mm diameter aperture further downstream the waste water jet enters in 2 cm distance from the nozzle into a separate beam catcher vacuum chamber where the by then supercooled liquid droplet stream freezes as slowly growing ice-needles onto a liquid nitrogen cooled cold trap, placed in 60 cm distance. With vacuum pumps of several 1000 L/s pumping speed and additional support by 10 000L/s cryotraps a vacuum of 10^{-6} mbar is sustained in the main

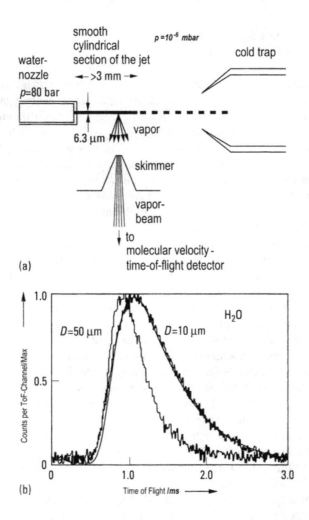

Fig. 4 Molecular beam time of flight spectroscopy of liquid microjet evaporation. **a** The microbeam setup shows the high-pressure nozzle, the free jet, and a beam-catcher cold trap for removal of the liquid jet. Part of the emerging vapor is extracted through an aperture in a conically shaped skimmer device. For velocity analysis a rotating disk with narrow slits is chopping the sampled molecular beam into short bursts, and molecules are detected in the ionizer of a mass spectrometer located at the end of a 0.81 m drift tube. **b** Two time of flight spectra of the velocity distribution of evaporated water from a thin 10 μm liquid jet and from a wider 50 μm jet, respectively, show a narrowed supersonic velocity distribution for the jet diameter larger than the vapor mean free path (D = 50 μm > l_{H_2O}) and, a broader Maxwell distribution (fitted by the smooth line) for collision free vapor expansion from 10 μm wide liquid surface into vacuum (figure reproduced from Ref. [3], Fig. 2)

chamber, surrounding the experimental surface probing region on the initial section of the intact, contiguous, 3 mm long liquid filament. For the molecular beam analysis of the vapor a sharp cone "skimmer" with ~ 50 μm entrance aperture, in two to five millimeters distance at right angle from the liquid jet propagation direction, is sampling a very small fraction of the radially evaporating, intense gas stream of water molecules. Not shown here in detail, these pass subsequently through a narrow slit in a rapidly rotating chopper wheel for producing short molecular beam bunches of 20 μs duration, and spread out in time in a drift tube on a vacuum flight path with 80 cm length. At the end of the drift region the dispersed molecules reach the electron bombardment ionizer region of a mass spectrometer placed in ultrahigh vacuum at pressures near 10^{-11} mbar for lowering the diffuse spurious gases background. By single ion counting and recording the time of ion arrival, time-of-flight spectra for the vapor molecule velocity distribution are accumulated, shown in Fig. 4b.

The two TOF spectra shown in Fig. 4b were obtained for two different sizes of liquid microjet nozzle diameter, one for $d_N = 50$ μm (which is $\gg l_{H_2O}$) and the other for $d_N = 10$ μm, respectively. The jet originating from the 10 μm pinhole nozzle is contracting to $D = 6.3$ μm diameter in the free surface flow region. The velocity distribution of the evaporated water from this narrower jet surface with free evaporation Knudsen conditions $D < l_{H_2O}$ is very close to the theoretical Maxwell distribution which is plotted also in Fig. 4b, as a smooth thin line superimposed to the "$d_N = 10$ μm" experimental spectrum. In contrast, the jet with the larger 50 μm nozzle, as an example for $D \gg l_{H_2O}$, shows a considerably narrower velocity distribution characteristic to ongoing supersonic jet expansion, driven by multiple molecular collisions in the early phase of vapor expansion. Thus, Fig. 4b gives the experimental proof that the free vacuum surface of liquid water with temperature above 0 °C can be prepared as high vacuum microjet surfaces when the jet diameter is smaller than the mean free path of the nascent vapor [2].

Numerical fitting of a Maxwell velocity distribution function to the measured distribution, in addition, yields the source temperature for the water vapor which is found to be a few degrees lower than the temperature of the nozzle and within expectations for a jet evaporational cooling model [2]. Finding a Maxwell distribution for the nascent vapor, at large, is in agreement with expectations for molecules which have to overcome a binding energy potential step barrier, i.e. the heat of evaporation, when moving from the liquid into the free vacuum space. Considering a thermal energy Boltzmann distribution in three independent cartesian coordinates, only the distribution component in the one coordinate perpendicular to the surface will be affected. Molecules will loose here the binding energy E_B in transit through the surface dividing plane, leaving a Boltzmann distribution with reduced intensity, however, with identical temperature T, as is easily seen by considering the mathematical separation formula: $\exp[(\frac{1}{2}mv^2 - E_B)/kT] = \exp(-E_B/kT)\exp(\frac{1}{2}mv^2/kT)$. This noteworthy finding that the vapor temperature, in spite of the evaporation energy loss of molecules, is identical to the liquid temperature, was an often disputed, surprising fact, although it had been published as early as in 1920 in a reply of Otto Stern to comments on his earliest Molecular Beam velocity distribution measurement [5].

In an additional upscaling experiment with the apparatus described in Fig. 4 we tried, in vain, to observe a direct evaporation of water dimer clusters from the surface of a microjet. Dimers are well known to occur in water vapor nozzle beam jets. In a mass spectrometer ionization source the dimers fractionize into $H-H_2O^+$ ions and are expected to appear at mass 19, next to the by far dominant water mass peak at mass 18. Actually, a faint, distinct signal peak could be observed at mass 19, 10^{-3} times smaller than the water monomer mass peak. However, when signal averaging the time-of flight spectrum over tens of hours at this purported peak the velocity distribution was exactly identical to the monomer peak. So it was identified to have come from the small fraction of deuterium and O^{17} atoms in HDO-water, and not from $(H_2O)_2$ water dimers with a mass of being twice the mass of monomer water that would result in average Maxwell distribution velocity smaller by a factor of $\sqrt{2}$. In conclusion, the water dimer fraction in evaporation was found to be smaller as least by a factor of five than the natural deuterium plus O^{17} abundance in hydrogen atom [2, 3, 6].

In continuing the search for direct dimer evaporation from liquid surfaces, carboxylic acids were studied which are known to form strongly bound dimers in a double hydrogen bridge structure COOH⊃HOOC-R, with binding energies in the order of 0.3 eV [6]. This is several times stronger than the water dimer hydrogen bond, and acetic acid microjets were found to emit large fractions, of $\geq 30\%$, of the vapor as dimers. The vapor velocity distribution measurements of the monomer species and of the dimer species, shown in the TOF spectra Fig. 5a, b, respectively, for liquid acetic acid, CH_3COOH, however, show yet another, very unexpected, phenomenon: When evaluating the measured velocity distribution by fitting a theoretical distribution function, the monomer distribution, Fig. 5a is very well fitted by a Maxwellian distribution function with a source temperature of ~252 K, well in the expected range of cooling for a vacuum microjet surface temperature. The acetic acid dimer velocity distribution, Fig. 5b, in contrast, can be fitted only by a slightly supersonic "floating Maxwellian" function representing the narrower half-width-spread by a Mach number of 3–4, and yielding a total dimer molecular beam enthalpy equivalent to a dimer molecules component apparent source temperature of 365 K for the liquid surface. This is 100 K higher than the apparent monomer source surface temperature and clearly above any error bar margins [6]. This anomaly in dimer source enthalpy of liquid surface vapor sources is further confirmed by vapor velocity measurements on a liquid jet of a mixture of 20% ethanoic acid in water, shown in Fig. 5c, d. The monomer distributions of the H_2O vapor component and of the CH_3COOH monomers, both displayed in Fig. 5c, are well fitted by simple Maxwellian function, with practically identical liquid surface source temperatures, 281 versus 275 K, although the average velocities of the two components differ by a factor of two, as to be expected for the mass ratio difference of 60:18 in molecular weight. This fitting result is also a proof of completely interaction-free, collisionless, vapor propagation of the two distinct mass components which show not any onset of dragging by the second component, known as the familiar seeded beams effect in more dense gas jet expansions. The dimer velocity component in this evaporating liquid mixture, in Fig. 5d, shows a slightly narrowed supersonic-like floating Maxwellian distribution

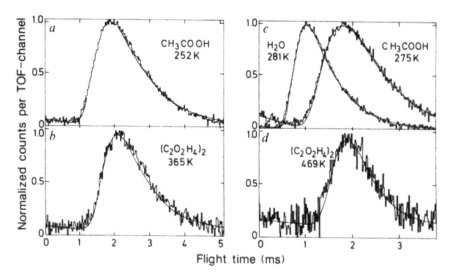

Fig. 5 Observation of non-equilibrium evaporation of dimer molecules of pure carboxylic acid and for a mixture of 20% CHCOOH in water. Temperatures shown are the (surface) source temperature calculated from the measured molecular beams time-of-flight velocity distributions. Dimers of acetic acid (Figs. 5b, d) show 100–200 K higher apparent source temperatures than the simultaneously evaporating monomers acetic acid (a) and for, both, H_2O and acetic acid monomers of the mixed solution (**c**) (figure reproduced from Ref. [6], Fig. 2)

with much higher apparent source temperature of 469 K, i.e., here the dimer source temperature is 200 K above the liquid surface temperature measured by the monomer vapor components emerging from the same liquid jet surface [6]. This astonishing anomaly is rising very interesting, far ranging questions about detailed microscopic balance of the liquid-evaporation/gas-phase-condensation process which requires the numbers of gas molecules evaporating per second and the number of molecules condensing must be in stationary equilibrium, as are to be their respective temperatures. As far as is known in statistical mechanics, or at least as far as is found in standard text books [7, 8], both, the distributions of evaporating molecules and the distribution of condensing molecules minus the distribution of molecules reflected from the liquid surface are all Maxwellian distributions with identical temperatures.

Although molecular liquid evaporation simulations for this phenomenon are not yet available, it was possible to give a semi-microscopic, intuitive model explanation for the observed velocity anomaly in dimer evaporation [6]: When the carboxylic dimers have formed in the liquid phase the two hydrogen bonding sites (at the O and at HO) of the, say, acetic acid COOH groups are crosswise saturated toward the adjacent dimer molecule. Then it may be reasonable to assume, that the two outward-pointing hydrophobic CH_3 groups act like a non-wetting nearly spherical inclusion in a water bubble or within a bubble of single hydrogen-bond, water molecules and monomerically dissolved acetic acid. When these preformed dimer inclusions approach the surface during the liquid evaporation process the bubble in this model

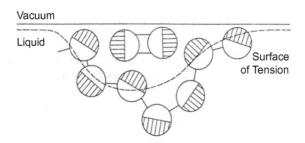

Fig. 6 Tentative surface tension model for the extra energy observed in dimer evaporation. For acetic acid CH_3COOH–$HOOCCH_3$ double hydrogen bonds form dimer inclusions in a liquid. The two CH_3 groups are pointing outwards and produce a hydrophobic dimer surface. The hydrophobic dimers are ejected by surface tension with an energy proportional to the released surface tension energy of the bubble hemisphere: $E \approx \sigma \pi R_2$. Different surface tensions of acetic acid ($\sigma = 1.7$ meV/$Å_2$) and of 20% acetic acid in water ($\sigma = 3.1$ meV/$Å_2$) explain readily the observed excess energies between dimer apparent source temperatures for the two liquids (figure reproduced from Ref. [6], Fig. 3)

picture will burst and the surface tension energy of the bubble will be released when the cavity is stretching out straight at the surface, transferring its freed energy like a stretched 'trampoline' to the outward moving dimer. This process is depicted in a cartoon drawing in Fig. 6. Assuming the hydrophobic dimer is sitting in a hemispherical cavity with radius of the average radius of the dimer molecule the change in surface area from hemispheric to plane, flat surface is just equal to the cross section area of the dimer molecule. With the macroscopically known surface tension value σ the energy gain for straightening out the Surface of Tension can then be calculated as $E_{excess} \approx \sigma \pi R^2$. For tabulated [9] macroscopic surface tension values of $\sigma_{Acetic} = 1.7$ meV $Å^{-2}$ for pure acetic acid at 0 °C and of $\sigma_{20\%Ac} = 3.1$ meV $Å^{-2}$ for a mixture of water with 20% acetic acid, it is obvious to expect a factor of two larger excess energy in the dimer evaporation from the water/acetic acid mixture, in agreement with the actually observed change of "surface excess temperature" from a value of 100 K in Fig. 5b to 200 K excess in Fig. 5d. Further measurements of dimer evaporation from formic acid ($\sigma = 2.7$ meV $Å^{-2}$, $T_{excess} = 150$ K) and for propionic acid confirm this model prediction of linear correlation of experimental excess dimer temperatures with the respective surface tension values [6]. The formal change of surface area calculated from the actually observed experimental value of 100 K for absolute cavity energy release in this crude model gives a value of $\pi R^2 \approx 7$ $Å^2$ model-effective geometrical cavity cross section area of the dimer $(CH_3COOH)^2$ molecule.

There are clearly many further unanswered phenomena left to study in evaporation of liquids, and examples of actually ongoing work on liquid surface molecular collision processes will be presented in the here following contribution, by G. Nathanson, to this Otto-Stern-Fest collection of talks.

4   Photoelectron Spectroscopy of Liquid Water

The free vacuum microjet surface, equally well, is suited for electron scattering investigations and for photoelectron emission studies of liquid water, and for a majority
of volatile liquid solvents in use in chemistry [10]. In particular, it has shown to be of
appreciable value for providing an experimental data base of electronic structure of
liquid aqueous solutions with widespread use for basic chemistry, electrochemistry,
and for some studies on biological substances.

In addition to providing vacuum compatibility, here, the fast-flowing liquid jet is
alleviating some of the notorious surface charging problems associated with photoemission from non-conducting materials. Charged speckles of the insulating liquid
surface are just washed away, instantly, with the microjet streaming speed of several
ten to hundreds of meters per second. Also, radiation heating effects and surface
damage by intense photon beams are reduced by orders of magnitude by the rapid
target replacement in the quasi-stationary surface of the microjet. Drawbacks and
newly arising problems of the liquid jet method are: electrochemical double layer
potentials build up readily near the liquid surface such as the "Stern Layer", for
example, describing in greater detail electrical polarization of molecules in the
surface layers of electrolytes; the moving liquid may be charged up dramatically
by electrokinetic phenomena related to the internal Zeta-potential value and of the
herewith associated Debye Layer thickness of the investigated liquid; in modestly
well conducting electrolytic liquids external superimposed electric fields can charge
up the liquid tip by current flow through the liquid filament column, leading to time
dependent surface charges interdependent with the droplet decay times [3, 10]. Thus,
additional measures had to be worked out for stabilizing, over the time of a photoelectron spectrum record, the electrical surface potential of poorly conducting liquid
jets in order to get meaningful absolute reference potential values for measurements
of photoelectron orbital binding energies from microjet photoemission spectra of
aqueous solutions.

The principal construction scheme of the microjet photoelectron spectroscopy
apparatus [11], shown in Fig. 7, resembles the earlier liquid jet free evaporation
molecular beam sampling apparatus that was given in Fig. 1. With the extension
for a photon beam directed onto the microjet on a third experimental axis, and,
after replacement of the previously used molecular beam time-of-flight detector by a
photoelectron hemispherical analyzer for the energy analysis, emitted photoelectrons
are detected at right angle with respect to the incoming photon beam and at right
angle with respect to the liquid direction. The UV and soft X-ray radiation used in
photoelectron spectroscopy are strongly absorbed by gases and need the microjet
vacuum for being able to penetrate the vapor shroud and to reach the liquid water
surface. The mean free path of electrons in gases is one order of magnitude larger
than the previously discussed free path for molecules.

For exploratory development and proofs of the technology the equipment was
tested initially with a Helium-I lamp laboratory radiation source for photon energy
$h\nu = 21.22$ eV (lambda = 58.43 nm). And after optimization for jet and for charging

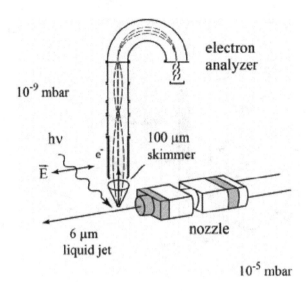

Fig. 7 Photoelectron spectroscopy on a free vacuum surface of a liquid microjet of water. A focused beam of photons with defined energy $E_{ph} = h\nu$, ranging from vacuum ultraviolet radiation to soft X-rays, is directed onto the microjet surface. Emitted photoelectrons are transferred into an electron energy analyzer with hemispherical electric deflection field for photoelectron spectra recording. The electron "vertical" binding energy E_B is determining the measured electron kinetic energy $E_{kin} = h\nu - E_B$ (figure reproduced from Ref. [11], Fig. 18)

stability, it was providing new photoelectron binding energy spectra for the three outer valence band electrons of pure liquid water as well as for some solvents such as ethanol or gasoline, and early data for solvated halogen ions in water solutions [3, 11]. Soon after, the PES microjet apparatus could be moved to 3rd generation Synchrotron tunable radiation sources (such as Bessy II) becoming available at the end of the 1990's. These yielded for microjets photoelectron spectroscopy studies sufficiently high radiation intensities of larger than 10^{16} monochromatized photons per square centimeter, focused onto a perfectly microjet-suited tiny spot of size <100 to <10 μm [10]. The Synchrotron radiation beam outlet port, not shown here in detail in Fig. 7, is protected by a series of several narrow collimator plus vacuum pump stages, needed to separate and to protect the Synchrotron storage ring ultrahigh vacuum region, at 10^{-10} mbar, from the water vapor loaded microjet surface intersection region.

A representative, typical set of energy resolved photoelectron spectra is shown in Fig. 8. It was obtained with Synchrotron radiation photons at $h\nu = 100$ eV for salt solutions of the diatomic alkali-halide salt series CsI, KI, NaI and LiI in water [13]. The photoelectron spectrum, in a simplified point of view, is imaging the electron density of states populations in the liquid electrolyte into an energy distribution spectrum of emitted electrons. The electron orbital binding energies are determined by the difference of the incident photon energy and the measured kinetic energy of emitted photoelectrons: $E_{bind} = h\nu - E_{kin}$. For the microjet spectra in Fig. 8, in molar concentrations of 2 m to 3 m, one salt molecule is dissolved in 20 molecules

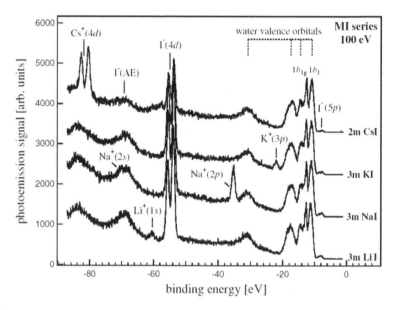

Fig. 8 Photoelectron spectra of alkali-halide salts in aqueous solution showing the valence electron states of liquid water, of the I⁻ ion and of the anions Li^+, Na^+, K^+, and Cs^+ (figure reproduced from Ref. [13], Fig. 2)

of water. At binding energies higher than the ionization onset threshold for water, near 10 eV at the right hand side the spectra of Fig. 8, a progression of three peaks is visible which are resulting from ionization of the outermost valence orbitals of liquid water, the states designated $1b_1$, $3a_1$, $1b_2$ of the water molecule and a fourth, broader peak, the inner valence orbital peak $2a_1$ at 32 eV binding energy. Energy level assignments and orbital energies for the isolated water molecule are depicted in Fig. 9, together with the correlation diagram for the origin of the H_2O hybridized orbitals from states of the separate oxygen and hydrogen atoms [11]. In addition to the 4 valence orbitals, the water molecule has one K-shell electron state, $1a_1$, at ~540 eV binding energy, to be seen later on in spectra obtained at higher photon energies. The experimental liquid water valence orbital peak energy position assignments are indicated above the valence spectrum by the horizontal scale bar in Fig. 8. In addition, in between the liquid water valence spectrum peaks, narrower faint water gas phase photoelectron peaks are visible which are caused by the vapor cloud surrounding the microjet. The most prominent gas phase spike here is designated by its gas-phase peak assignment $1b_{1g}$ on the upper spectrum in Fig. 8 and it is shifted by approximately 1.5 eV "gas-liquid shift" with respect to the liquid phase $1b_1$ feature. During measurements, the water gas phase peak is a very helpful reference calibration point in undergoing microjet spectra evaluations. It is averaging over the electric field in the immediate surroundings of the liquid jet and, thereby, also can be used as an indicator of unintentional or unnoticed jet charging. When the potential at the jet surface is differing from the grounded chamber wall the resulting electric field

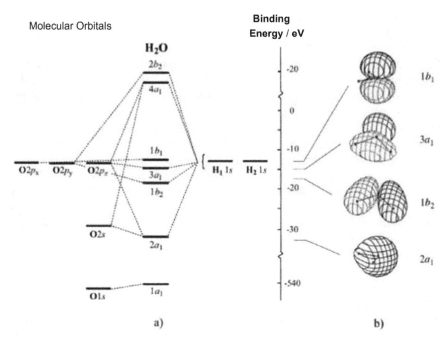

Fig. 9 a Molecular orbitals energy level diagram of the gas-phase water molecule. Intermittent diagonal lines indicate contributions of atomic orbital states of the separate H-atoms and O atom to the H_2O molecular orbitals. **b** Electron density contours for the highest occupied valence shell orbitals $1b_1$, $3a_1$, $1b_1$, and the "inner valence" orbital $2a_1$ (figure reproduced from Ref. [11], Fig. 20)

gradient is causing a clearly noticeable, field induced broadening of the gas phase $1b_{1g}$ peak and in a change of relative energy difference between liquid peak maximum and the gas phase peak.

The remarkable peak broadening of all liquid phase water valence electron energy levels with respect to the gas phase photoelectron spectrum, by amounts of the order of 3 eV peak half width, was surprising when found in the first measurements of liquid water photoelectron spectra [10]. It appears to be associated with the multiple and rapidly varying hydrogen bonding and thus seems to reflect the heterogeneous environment of liquid water molecules [11]. However, no simple broadening explanation is visible from water simulation calculations. The numerous salt ion peaks appearing in the electrolyte spectra are notably narrower, of an order of 1.1–1.5 eV, what is probably caused by the more rigid, stable hydration shells forming in ion solvation. The I^- ion in aqueous solution, for example, shows two marked photoemission peak doublets, a weak peak at 7.7/8.8 eV originating from ionization of the outermost electrons in the I^- (5p) orbital of iodine and a strong (here resonance enhanced) feature for the photoelectron emission from a lower orbital, I^- ($4d_{5/2,3/2}$) at ionization energy 53.8/55.5 eV. A weaker emission structure, I^-(AE), appearing in the spectra at the position near 70 eV on the binding energy scale, was identified as Auger electron emission occurring when an electron hole created in the I^-(4d) shell by direct photoemission is filled up shortly after by one $I^-(5p)$ outer orbital electron, at the simultaneous emission of a second 5p electron with the excess energy difference between the $I^-(5d)$ binding energy and the 5d $\rightarrow$ 4d transition energy. Constant emission energy at variations of the photon energy is the experimentally easy to confirm signature for this process. The alkali-anion peak structures in the spectra allow the experimental determination of the vertical ionization energies for the outer shell electrons of the series $Li^+(1\ s)$, $Na^+(2p)$ and $K^+(3p)$ with values of $E_{aq}^{PES} = 60.4$ eV for Li^+, 35.4 eV for Na^+,22.2 eV for K^+, respectively [11, 13]. For Cs^+ the vertical ionization energy peak, estimated at about 15 eV, is buried in the intense liquid water valence structure peaks. The vertical binding energy values, thus obtained, are of great relevance for theoretical description of chemical processes in aqueous solution. They differ from the chemical solvation enthalpy G^0 of equilibrium caloric experiments by the relaxation energies which arise by the rapid formation and realignment of the solvation shell of solvent molecules around a newly formed solvate ion species, and therefore allow an experimental determination of the important reorganization energies in electrochemical processes in polar solvents. Noteworthy, a comparison with the already known and tabulated gas phase ionization energies of the alkali and halide ions [9] shows the solvated states are changing by remarkably large amounts of energies: for I^- gas phase ions the ionization energy is 3.1 eV and shows a gas-liquid shift to 7.7 eV in the aqueous solution experiment shown in Fig. 8. This shift is opposite in sign to the gas-liquid shift of about 12 eV for Na^+ ions, decreasing from the gas phase ionization energy of 47.3 eV to lower electron binding energy of 35.4 eV in the liquid.

Whereas direct simulation of the peak solvation energy shifts is proofing to be demanding and time expensive, Max Born in 1922 had already proposed an elegant, very descriptive dielectric cavity (DCS) model for ion solvation which is sometimes still in use in phenomenological description of electrolytic fluids. Here the ion is thought to sit inside an empty spherical cavity inside a continuous dielectric medium with dielectric constant $\varepsilon\ \varepsilon_0$ different from the vacuum dielectric value ε_0. Using the electrostatic Maxwell equation for calculating the energy of a sphere with charge Z and radius R in an infinite dielectric medium and comparing versus vacuum with dielectric constant $\varepsilon = 1$ he obtains the formula: $\Delta E_{solvation} = (Z^2\ e^2/8\ \pi\ \varepsilon_0\ R)$ $1/(1 - 1/\varepsilon)$ for the solvation energy of an ion with charge Z and ion radius R, giving reasonably good numerical results for positive monatomic ions in water [10]. In the case of I^- photodetachment Born-model results are far less accurate, simply because at final state with zero charge, the reorganization time of the dielectric water is difficult to include appropriately. The comparably small gas-liquid shift of 1.5 eV, only, in the ionization energy of the neutral water orbitals may also be rationalized using educated guess estimates for time-dependent dielectric constants in the DCS model. For water the static value of the dielectric constant is $\varepsilon \approx 80$. This constant is decreasing for time varying electric fields. On extremely short experimental time scales, such as the time period of light, the dielectric constant ε approaches the square value of the optical diffraction index n of water (n = $\sqrt{\varepsilon} \approx 1.4$) according to Maxwell's relation between the velocity of light and dielectric constant and the fact that diffraction indices are deriving from ratios of velocities of light. Therefore, for water dielectric reorganization on photoelectron emission with time spans of 10^{-15} s before the electron is distanced, the more decent estimate for the effective, dynamic dielectric response constant to the suddenly formed ion is likely to be closer to a value of $\varepsilon_{fs} \approx 2$. In accord with this consideration, using the "optical" value $\varepsilon_{fs} \approx 2$ in Born's DCS formula yields a gas-liquid shift of the order of 1.5 eV for newly formed H_2O^+ ions vertical ionization energies, in fairly good agreement with the observed shifts in the present water PES spectrum measurements, Fig. 8. The neutral liquid water had no time yet to respond to the appearance of the photoionized H_2O^+ ion in the neutral water medium. The DCM asymptotic model is thus giving at hands an intuitive, and roughly predictive formula which should be useful when discussing the far more accurate, but, by far less transparent results of realistic computer simulations of liquids.

The measurements in a wide range of ion concentrations and for different counterions show a remarkable independence of the ionization energies which are not changing within the current experimental accuracy of 30 meV (equivalent 3 kJ/mol in more familiar caloric units). These vertical ionization energies, also, are not varying for different locations in the electrolyte, independent of whether the ionization takes place near the surface or in greater depth in bulk phase environment, as is demonstrated by a series of measurements on 2 m solutions of NaI with photon energies being changed between 200 and 1000 eV, in Fig. 10. Here is made use of the fact

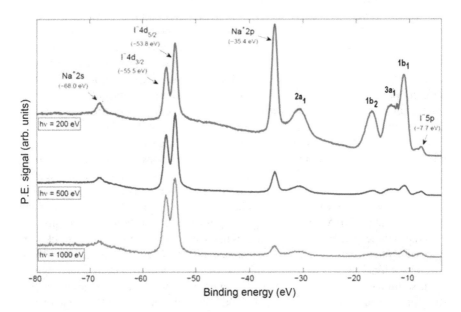

Fig. 10 Photoelectron spectrum of 4m NaI aqueous solution, taken at three different photon energies 200, 500, and 1000 eV. With increasing kinetic energy of photoelectrons deeper bulk regions of the aqueous solution are probed. The minimum escape depth is 2 or 3 water molecule diameters at 200 eV electron kinetic energy, and the probing depth increases to a layer thickness of 10–12 H_2O-diameters for 1000 eV photoelectron energy (figure reproduced from Ref. [14], Fig. 8)

that for photoelectrons with different kinetic energies the escape depth for electrons increases with increasing energy [14]. The thickness of the water layer probed increases, approximately, from 3 molecular layers at 150 eV (≈ 12 Å) to a layer thickness of 12–15 molecular diameters of water at 950 eV kinetic energy. The observed photoelectron signal for a given depth and given initial energy is exponentially attenuated when the photoelectrons penetrate to the vacuum space over larger distances below the surface, whereas the photoelectron energy is not altered much by electronic stopping power on these comparably short average escape path lengths. This is confirmed by the observation, that the NaI photoelectron peak shapes are not broadening with increasing photon energy, although, due to the associated higher kinetic electron energies, photoelectrons are probed for significantly greater sampling depth [14].

It may be noticed that the photoelectron spectra recorded in Fig. 10 are very smooth and show almost negligibly small statistical counting noise in comparison to the earlier measurements of Fig. 8. This is the result of continuing experimental improvements of the synchrotron radiation photon beamline and of the water photoelectron spectroscopy apparatus. The photon intensity is higher and it is focused to a smaller spot size, illuminating fully a 15 μm liquid jet. In the liquid water valence peak features, therefore, the gas phase peak $1b_{1g}$ has vanished almost completely, and is visible only as the very weak spike residue between the $1b_1$ and the $3a_1$ liquid water photoelectron peaks. The three spectra are shown here with intensities normalized to the strong I^- 4d peak feature. The wide variation of all other relative peak intensities in between the three different spectrum records is caused, primarily, by changes in photoemission cross sections for different incident photon energies.

Basically, however, the photoelectron peak intensities are proportional to absolute concentrations of molecules in the liquid target probe. Depth profile probing with different photon energies, hence, allows, also, quantitative studies of concentration changes near the liquid-gas surface of solutions. Listed ionization cross section data of reasonable computational accuracy are available by NIST [12]. Using these and the experimental apparatus functions for collection and transmission of photoelectrons it is straightforward to evaluate absolute ion concentrations for Na^+ and for I^- from the PES measurements [14]. The thus obtained I^-/Na^+ ratios over the photoelectron kinetic energy range from $E_{kin} = 100$–1000 eV, are plotted versus the electron kinetic energy, in Fig. 11. They show a clear enhancement by almost a factor of two in favor of I^- anions near 200 eV kinetic energy, sampling the composition within the first two or perhaps three water molecule diameters in the uppermost layers of the liquid. With increasing photoelectron energy the evaluated ion ratios decrease and, above 400 eV, are approaching asymptotically the value of one, expected for the bulk stoichiometry of the sodium iodide salt solution. In a computational molecular dynamics modelling study of 1.2 m alkali halide salt aqueous solutions (with 18 Na^+ and 18 halogen-anion molecules in a slab of 864 H_2O molecules in the numerical sample calculation) the surface enhancement for halogen anions concentrations with increasing anion radii had been studied theoretically, one decade earlier [15], and can here be compared with detailed experiments. Snapshots of ion and water molecules distributions from this molecular modelling study are reproduced as a side insert in Fig. 11, adjacent to

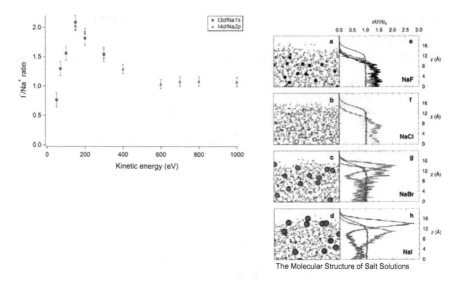

The Molecular Structure of Salt Solutions

Fig. 11 (At left) evaluated anion/cation intensity ratios at different photoelectron kinetic energies show a significant propensity of I⁻ ions over Na+ near the surface of the 4 m NaI aqueous solution. At electron kinetic energies larger 500 eV the measured ratio approaches the bulk solution ratio 1:1. (At right) snapshots of computer simulations of the molecular structure of alkali-halide salts in aqueous solution show a preference for Iodine ions to the liquid-gas surface. Evaluated anion-, cation- and, water molecule-densities plotted versus the z-position coordinate of the simulation slab (Fig. 10h) are showing in quantitative detail an enhancement of the (magenta) I⁻ anion concentration at the water surface and a tendency for immersion of Na+ cations (red) into the bulk aqueous solution (O-atoms:blue) (figure 10 adapted from Ref. 12, Fig. 14 and Ref. 15, Fig. 1)

the measurement, and illustrate the prevalence of I^- anions over the Na^+ cation near the model water surface [15].

5   Core Electron Spectroscopy of Protonation/Deprotanation in Aqueus Solution

High resolution photoelectron spectroscopy at here available energies up to 1000 eV provides, also, a useful, sensitive, tool for chemical-environment sensitive K-shell core electron spectroscopy in a number of low-Z atoms, such as C atoms, N, or O-atoms which are prevailing in solute organic molecules. Chemical environment induced shifts in K-shell photoelectron spectroscopy are well known since the earliest studies of Kai Siegbahn's group in the 1960's (on solid state probes), then coined as ESCA, the electron spectroscopy chemical analysis. A great advantage of core shell spectra is in identifying chemical changes near a single, specific, atom in a chemical compound.

In Figs. 12 and 13 it is illustrated how the pH value induced change by protonation

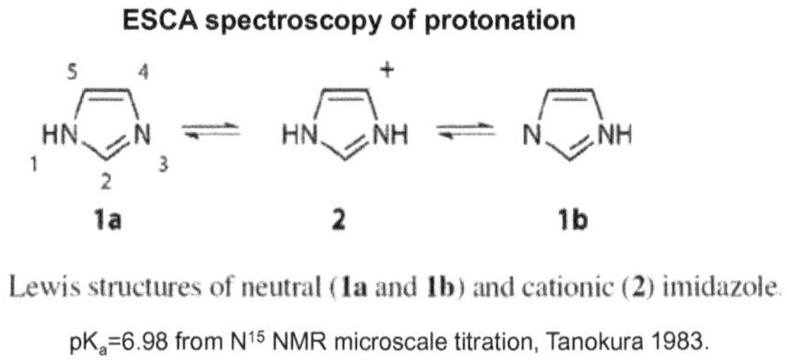

Fig. 12 Lewis structures of neutral (1a and 1b) and of cationic (2) imidazole. Known $pK_a = 6.98$ from N^{15} NMR- microscale titration (Tanokura 1983) (figure reproduced from Ref. [16], Fig. 1)

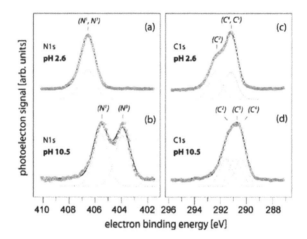

ESCA spectroscopy of protonation

Lewis structures of neutral (**1a** and **1b**) and cationic (**2**) imidazole.

$pK_a = 6.98$ from N^{15} NMR microscale titration, Tanokura 1983.

Fig. 13 Inner shell Nitrogen 1s (**a, b**) and Carbon 1s (**c, d**) photoelectron spectra show energy shifts for protonation and deprotonation of 2m imidazole aqueous solutions measured at pH 2.6 (cationic structure 2), and at pH 10.5 (neutral molecule 1a, 1b). The photon energy is 480 eV for the N1s measurement and 380 eV for C1s. Smooth intermittent lines show fitting results to the experimental spectra (red circles). In the protonated state the two nitrogen atoms in imidazole are indistinguishable (**a**) (figure reproduced from Ref. 16, Fig. 2)

can be traced on individual atoms of a solvated organic molecule [16]. Imidazole is a five atom ring compound, made of two N-atoms and three C-atoms, shown in the Lewis-structure representation of the molecule in Fig. 12. Protonation of the neutral dissolved imidazole molecule leads to two identical NH groups appear in the molecule, Fig. 12 structure (2), while the charge of the added proton is delocalized over the whole molecule. The pK value for imidazole protonation of $pK_a = 6.98$ had been determined by, isotopically enriched, N^{15} NMR microscale titration by Tanokura, in 1983 [16]. The 1 s core electron photoelectron spectra for the N1s and for the C1s core states in 2 m imidazole solutions for two pH values, in Fig. 13, show the effects of the transition from neutral to protonated molecules. At pH 10.5, in the lower row of spectra, the imidazole molecule is neutral and, as expected from the structure formulas 1a and 1b for the neutral compound, two separate N1 s energy levels are observed for the two different nitrogen atoms in the ring, one peak for the nitrogen atom N^1 where the hydrogen atom is attached and, separated from the first peak by 1.5 eV, a different peak N^3 for the second nitrogen in position 3 of the ring (indicated in the structure drawing in Fig. 12). When changing the pH value both peaks are decreasing and a third peak, in the position of the N1s single peak labeled (N^1, N^3) in the upper spectrum is growing up in intensity. In subsequent measurement with solutions of different pH values, not shown here for shortness, after crossing pK = 7 value the first double peaked structure is shrinking further in amplitude until the fully protonated imidazole solution at pH 2.6 shows only one single photoelectron peak for nitrogen, confirming theoretical chemistry results that here two pseudo-equivalent NH groups exist with the positive charge/electron hole-orbital distributed equally over the location of both N-atoms, and not a NH group and a distinct NH^+ is formed after the proton attachment. This transition of the binding sites of the two distinct N atoms in the neutral state to two identically bound nitrogens in the protonized ion is further reflected in the C1s peak structures of the three carbon atoms bound in the molecule's ring structure. For the neutral imidazole, Fig. 13d, at pH 10.5 a broadened carbon C1s photoelectron spectrum is observed which can be deconvoluted into 3 nearby lying C1s states with similar amplitudes, corresponding to the three different C/N neighborhood bonding configurations of the three carbon atoms in the positions 2, 5 and 4, respectively, indicated in the, Fig. 12, structure scheme drawings 1a and 1b. In the charged state, Fig. 13c at pH 2.6, however, these carbon C1s levels contract to two overlapping states in new positions on the energy scale, as shows the deconvolution of the peak structure into two standard width peaks. The stronger peak is attributed to, both, the C^4 and the C^5 atoms, which are now in identical neighborhoods, as expected for the protonated imidazole species. The deconvoluted core shell C1s peak amplitudes, individually image the stoichiometric ratios for different atoms in the photoelectron spectrum and, accordingly, the joint (C^4, C^5) carbon 1s peak shows twice the amplitude of the separate C^2 peak originating from bottom C-atom C^2 connecting the two identical NH groups in the protonated imidazole. In summary this shows, here titration of charged/neutral molecular states can be performed quantitatively, in stoichiometric precision.

In further detail, this 1s core level PES titration demonstrates, in addition, the very distinct methodical advantage of the exceptionally high intrinsic time resolution on the order of sub-femtoseconds, given for the photoemission process by the time scale for the removal of the fast electron from the parent atom. The protonation/deprotonation bond-making and bond-breaking processes in solution take place on time scales of 10^{-12} s. Thus, in the 1 s photoelectron spectra of solutions near the pK_a point always two distinct peaks for protonated and deprotonated species populations appear simultaneously. In contrast in the classical NMR microtitration procedure the averaging time is limited by the period of the absorbed resonance frequency, in an order of 10^{-8}–10^{-9} s. Therefore, averaging over many proton bond making-and-breaking cycles occurs in the NMR method which results in a frequency shift with weighted averaging over the two distinct states, only, without simultaneous separation of both levels.

A somewhat more complex case of chemical adsorption and reaction is investigated by photoelectron spectroscopy studies shown, in Fig. 14, for analysis of the details of carbon dioxide capture in industrially used solutions of Monoethanolamine ($HOC_2H_4NH_2$) for washing CO_2 from flue gas [17]. Known for more than a century, the chemical steps involved in the gas capture process have been extensively studied and characterized in great detail. 30% monoethanolamine (MEA) in aqueous solution has a CO_2 load capacity of 0.25 mol/L. The principal capture reaction is:

$$2\,MEA + CO_2 \rightarrow MEA - COO^- + MEA - H^+$$

Acid/base equilibria are:

$$MEA + H_2O \leftrightarrow MEA - H^+ + OH^- \quad pK_a = 9.55$$

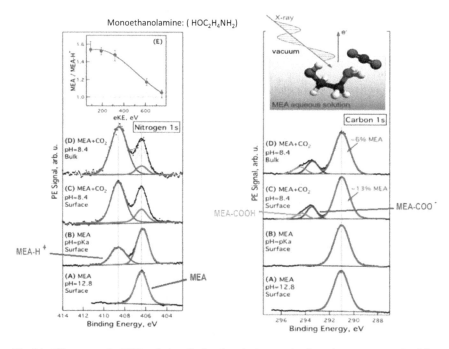

Fig. 14 CO_2 capture in MEA solution (Left column) nitrogen 1s photoelectron spectra for 4.9 m MEA (A, B) and for CO_2 treated MEA solutions (C, D) show varying contributions from MEA in its neutral form (red, 406.3 eV BE) and from protonated MEA (blue, 408.6 eV). Red intermittent lines represent carbamate reaction product contributions (see text). The ratio (E) of neutral and protonated MEA molecules changes as a function of depth in a solution of CO_2 loaded MEA at pH 8.4 (Right column) carbon 1s PES spectra for MEA (A, B) and CO_2 loaded MEA (C, D). In CO_2 treated MEA separate peaks from carbamate (low BE, purple) and carbamic acid (green, high BE) appear. Red labels on the peaks for BE=291 eV indicate the percentage of neutral MEA contribution (figure from Ref. [17], Figs. 1 and 2)

$$MEA - COOH + H_2O \leftrightarrow MEA - COO^- + H_3O^+ \quad pK_a \text{ unknown } (7 - 9?)$$

Photoelectron spectra for C1s and for N1s were taken for MEA-CO$_2$, gas loaded solutions (Fig. 14c, d) and for 30% MEA solution without gas load (Fig. 14a, b). pH values are adjusted to the technical working point, or to other values, when needed for analysis of details. At pH 12.8 the MEA in solution is completely neutral, and the N1s binding energy spectrum (A), at the bottom line of Fig. 14, shows a single nitrogen atom peak centered at $E_B = 406.4$ eV with half width 1.3 eV. The spectrum is a surface spectrum, taken at electron kinetic excess energy 90 eV. In (B) the pH value was adjusted to the (bulk) $pK_a = 9.5$. In this surface spectrum a second peak, smaller by a ratio 1.6, appears at 408.8 eV BE and is identified as the N1s signature of the protonated NH$_3^+$ group in the MEA-H$^+$ fraction. The ratio of the two peak amplitudes represents the quantitative ratio of the two MEA species in the locally probed region of the solution. The ratio was determined in a series of additional measurements with progressively higher photoelectron excess kinetic energies in a range up to 750 eV and shows, in plot (E) in the upper row of Fig. 14, the ratio is reducing from an excess 1.6:1 of neutral MEA-molecules near the surface to a 1:1 ratio for 750 eV electrons which originate in greater depth from the bulk phase of the liquid.

The two remaining N1s spectra (C) and (D) are taken for gas saturated MEA solution with a CO$_2$ load of 0.24 mol/L. These solutions change the original, equilibrium pH value from pK$_a$ to pH = 8.4 in the loaded state. The transformation of MEA to carbamate MEA-COO$^-$ and to carbamic acid MEA-COOH changes little in the N1s binding energy of the NH$_2$ group on the opposite end of the molecule. The peak appearing at the position of "neutral MEA" in these spectra is the superposition of unknown fractions of contributions from loaded and non-loaded MEA with two almost identical peak shapes. More can be learned from a consideration of C1s spectra shown, adjacent to respective N1s results, on the right-hand side of Fig. 14. For neutral MEA at pH 12.8, and for MEA at pH = 9.5 a single narrow C1s peak structure for all carbon atoms in the compound is observed at 291 eV binding energy in the spectra shown in (A) and (B), respectively. In the CO$_2$ saturated MEA solution at the surface (C) and for bulk solution (D) two new, distinctly visible C1s peaks arise from carbamate (MEA-COO$^-$) and for carbamic acid (MEA-COOH), with energies shifted to higher binding energy. From the intensity ratios of the carbamate and of the carbic acid peaks, at known pH-value of the solution, the first experimental determination of the previously uncertain pK$_a$ equilibrium with a pK$_a = 8.2$ is here obtained. Further evaluation of the C1s peak ratios, in combination with measured relations between neutral and protonated species from the MEA/MEA-H$^+$ ratios of the simultaneously recorded N1s peak spectra, eventually, yields an absolute concentration ratio of (MEA) over (MEA-COO$^-$ + MEA-COOH) of ~0.22 when probing the surface and of ~0.09 for probing the bulk. The carbamate products have a preference for moving into the bulk and MEA a tendency to be enriched on the surface. This provides a perfectly cooperative cycling support for the CO$_2$ trapping process at the interface and for the subsequent removal of carbamates into the bulk of the washing fluid [17].

6 Excited States of Water, Resonant Auger Spectroscopy of H_2O_{aq} and OH_{aq}^-

Liquid water is dissociating spontaneously into a very small fraction of H_3O^+ and OH^- ions, with far ranging consequences on the properties of aqueous solutions. The ions with concentrations of 10^{-7} mol/L in pure water, by far, are too small to be observed in photoelectron spectra. Thus, solutions of strong acids (1 m HCl, corresponding to pH = -1) and of strong bases ($LiOH_{aq}$), instead, are to be used for photoelectron spectroscopy of the self-ionization products of water [18]. In spectra, shown in Fig. 15, valence orbitals for H_3O^+ and for OH^- have been studied and show a weak perturbation of the dominant H_2O valence photoelectron structure, superimposed by faint photoelectron emission from H3O+, in Fig. 15a, and a small peak localized at the water ionization threshold with an OH^- ionization energy of 9.2 eV,

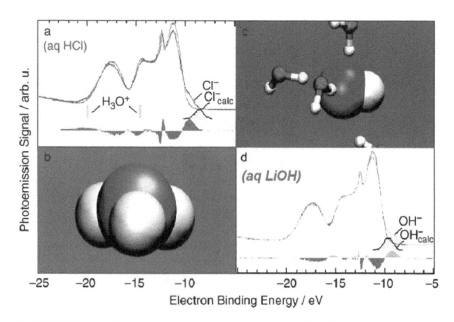

Fig. 15 $H_3O_{aq}^+$ and OH_{aq}^- valence photoelectron spectra obtained by difference measurements of 3m HCl aqueous solution with neat water (**a**) and of a 2mLiOH solution (d). The $H_3O_{aq}^+$ ion signature appears as a weak peak at 20.8 eV BE, in the tail region of the $2a_1$ inner valence orbital of water. The OH^- electron binding energy in solution is comparable to other halogen ions. Also, for comparison with the experiment, theoretically assigned ionization energy positions for $H_3O_{aq}^+$ and OH_{aq}^- are indicated by grey shaded features (figure reproduced from Ref. [18], Fig. 2)

for the LiOH solution in Fig. 15d, similar in magnitude to the previously considered photoionization energies of halogen anions. Separately taken, photoelectron spectra of pure water are here compared to the photoelectron spectra of the H^+ and OH^- in solutions and are used for extracting difference spectra shown in the lower part of Fig. 15a, d. Simulated spectra for calculated values for the OH^- and the Cl^- ionization potential are also shown, and are found to be in reasonable agreement with the measurement [18]. For the solvated proton different hydronium configurations were considered in the model calculation. The "Eigen"-like aqueous cation structure H_3O^+ was judged to agree best with the experimental ionization energies derived from the HCl solution spectra, as indicated by vertical bars for theory results, in Fig. (15a).

OH^- in the gas phase exists for a single negative ion ground state, only, and similarly to the anions of halogen atoms the attractive potential well is too shallow to support any exited electronic state. In contrast, in aqueous solutions the anions are embedded in liquid water in the additional polarization potential described by the dielectric-cavity/Born-model which increases the well depth and binding energy to about 10 eV, a value large enough to allow for the existence of an excited electron state, at a binding energy of the order of 1 or 2 eV below the ionization threshold. The optical transition to this excited state, a very strong s $\rightarrow$ p absorption line, was first observed in the 1930's in UV absorption spectra of I^-_{aq}. The phenomenon was called charge transfer to solvent (CTTS) and has long drawn the attention of spectroscopists because the CTTS states existed in liquids, only.

With the availability of narrow band, tunable soft X-ray synchrotron radiation it became possible, also, to explore this CTTS band by resonant excitation from an inner core level, and monitoring the resonance in the Auger electron emission spectrum. This has the advantage that the limitations of "classical" UV/VUV spectroscopy by the onset of the strong absorption of liquid water above 9 or 10 eV can be offset in the Auger method. Before testing with OH^-_{aq} in liquid water, we explored the technology first on Cl^- ions where the process is simpler and better known from previous optical VUV spectroscopy work [19]. An illustration of the different possible Auger excitation processes is shown by the three schemes drawn, in Fig. 16, for the Cl^- anion. Direct Auger electrons (1) are emitted following an excitation of a core hole vacancy by photoelectron emission. In a rapidly following step, within a few fs, the hole in the inner shell 2p-level is filled by an outer valence 3p electron and the gained energy is transferred to a second outer valence shell 3p electron which is emitted as the (LMM-) Auger electron. Spectator Auger electrons (2) are emitted after resonant excitation of a 2p electron into unoccupied levels "e1, e2, ..." of the solvated Cl^-_{aq} ion. The energy of the spectator-Auger electron is higher than for normal Auger electrons because the originally photon-excited electron is still present in the ion and increases the coulomb forces acting on the outgoing Auger electron. A third process may occur (3) in a shake down, transferring electron-energy of the resonantly populated levels to other internal states before the Auger electron is emitted, with the result of an additional change in the kinetic energy of the emitted spectator-Auger electron.

In a series of photoelectron spectra records at closely spaced energies of the incoming synchrotron radiation, tuned over the region of interest for expected 2p to CTTS transition, shown in Fig. 17, these discussed Auger phenomena can be

2p core-hole excitation of the chloride anion, Cl⁻ (aq)

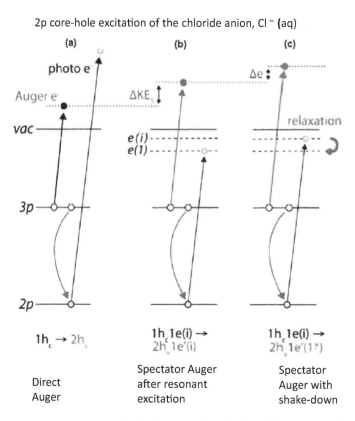

Fig. 16 Auger processes associated with 2p core-hole excitation of the chloride anion, Cl^-_{aq}, in aqueous solution. Negative halogen ions have no excited states in the gas phase. In aqueous solution one or several new excited states appear, called charge transfer to solvent states (CTTS). **a** Direct Auger process with emission of a photoelectron followed by emission of the Auger electron. **b, c** Spectator Auger emission after resonant excitation of one of the CTTS states e(1) − e(i). The spectator Auger electron kinetic energies are shifted due to electrostatic interaction by the presence of the resonance electron. **c** Occasionally, energy of the resonantly excited state is transferred to nearby internal states before the Auger electron is emitted (figure reproduced from Ref. [19], Fig. 5)

actually observed [19]. At the lowest photon energy, at 200 eV, the electron kinetic energy spectrum shows only the familiar peak structures of the liquid water valence bond states, and in addition, the small Cl^-_{aq} peak shoulder adjacent to the right-hand side of the water $1b_1$ peak. At the highest photon energy 204.8, the electron kinetic energy spectrum of the water valence structure photoelectron spectrum is shifted toward 4.8 eV higher kinetic energies in accordance with the higher photon energy. In addition, the 204.8 eV spectrum shows the strong, fully developed, regular Auger peak for Cl^-_{aq} LMM Auger emission with indicated doublet splitting n, n̲. The LMM Auger peak starts to develop at photon energies larger than the $2p_{1/2,3/2}$ level ionization energy of 201 eV. Its signature is the constant kinetic electron energy, independent from the incident photon energy. Most importantly, however, in the intermediate

Fig. 17 Photoelectron and Auger electron spectra for Cl^-_{aq} measured at photon energies near the detachment energy for 2p orbital electrons. A 3m LiCl aqueous solution liquid jet is used. Normal Auger peaks for the Cl^-_{aq} $2p_{3/2}$ (**n**) and $2p_{1/2}$ (**n_**) LMM process (see Fig. 16a) occur at constant kinetic energy. Emitted photoelectron peak positions for water move with increasing photon energy. Resonant absorption transitions into unoccupied CTTS states of Cl^-_{aq} are readily observable at the resonant Auger line positions **1_, 2, 2_, 3_, 4**. Resonance features **2, 4** are attributed to transitions originating from $p_{3/2}$, **1_, 2_** and **3_** from $p_{1/2}$ Auger resonances in CTTS state (figure reproduced from Ref. 19, Fig. 2)

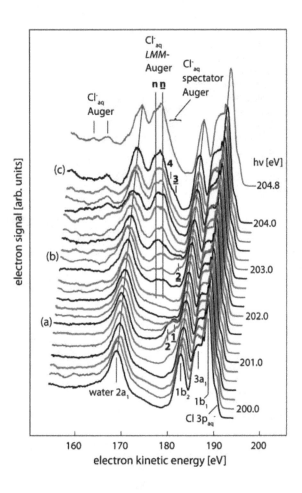

region of scanned photon excitation energies some photoelectron spectra show small intensity bumps appearing only in very narrow excitation energy ranges ≤ 0.3 eV full half width, which originate from the searched-for resonant excitation of Cl^- unoccupied level states. The resonance Auger features are identified in Fig. 17 by the numbers $\underline{1}$, $\underline{2}$, $\underline{3}$ and 2 or 4, pointing to the observed features at three different excitation photon energies where resonances could be detected. The region of the $(2, \underline{1})$ Cl^-_{aq} spectator-Auger peaks group is drawn enlarged in Fig. 18, showing the signal evolution as a function of photon energy. The blue peaks are Gaussian fits of 2 and $\underline{1}$ to the experimental kinetic energy spectra envelopes. After analogous evaluation of all other observed spectator-Auger resonances an energy level diagram of these newly identified unoccupied electronic levels can be constructed and is shown in Fig. 19 together with the also determined absolute energy level values for the occupied electron orbital states 3p and $2p_{1/2}$, $2p_{3/2}$ which were obtained simultaneously from the measured photoelectron spectra of Cl^-_{aq}. The Cl^- excited states orbitals in aqueous solutions are here found at binding energies of 2.5 and

Fig. 18 Enlarged region of Cl_{aq}^- photoelectron spectra in Fig. 17, showing the signal evolution of spectator Auger-electron peaks **2** and **1_** at photon energies between 200.4 eV and 201.4 eV. Blue peaks are Gaussian fits to structures **2** and **1_** in the experimental photoelectron spectra (red) (figure reproduced from Ref. 19, Fig. 3)

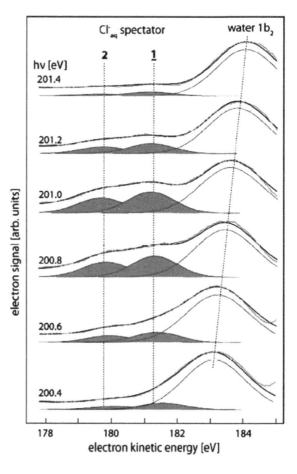

Fig. 19 Experimental CTTS states of Cl-**aq** and electron binding energies for the 3p and 2p states of the negative chlorine ion in 3 m LiCl aqueous solution (figure reproduced from Ref. 19, Fig. 6)

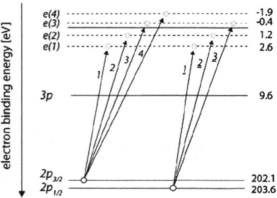

of 1.3 eV, and two further, antibound state resonances were identified, appearing slightly above the vacuum level, at −0.4 and −1.9 eV, respectively [19].

With the expertise acquired in resonant spectroscopy of CTTS states in the simple, spherical CI_{aq}^- anion, we return to the OH_{aq}^- with a study to find the CTTS state here in resonant excitation from the 1 s core state of the O-atom in OH_{aq}^- which is the only inner shell state available in this anion entity [20]. Fig. 20 shows electron kinetic energy spectra obtained in a narrow scan range of photon energies near the expected O1s to CTTS transition. In a series of preceding photoelectron experiments the OH_{aq}^- O1s binding energy had been determined to be 536.0 eV (shown in Fig. 21, in a combined experimental energy level diagram for liquid H_2O and aqueous OH−). In search for the 1s resonance of OH_{aq}^- near the continuum threshold, kinetic energy electron spectra are shown, in Fig. 20, for 4 m NaOH solutions (red lines) and for

Fig. 20 4 m NaOH aqueous solution photoelectron spectra, showing resonant Auger-electron spectra (**b**, **c**) at 532.2 and 532.8 eV, and (ultrafast) intermolecular coulombic decay of OH_{aq}^-. The photon energy is scanned near the ionization threshold for the oxygen O1s inner shell orbital. In addition, photoelectron spectra of pure water are recorded (blue) for reference. Resonance Auger electron peaks 2, 3, and 4 reveal the existence of a very fast energy transfer process from excited OH− to the H_2O solvent. The small peak at highest kinetic energy, far right, arises from O1s ionization by spurious, second harmonics photons with energy 2 hν, and, provides a method for highly accurate absolute energy calibration (figure reproduced from Ref. [20], Fig. 1)

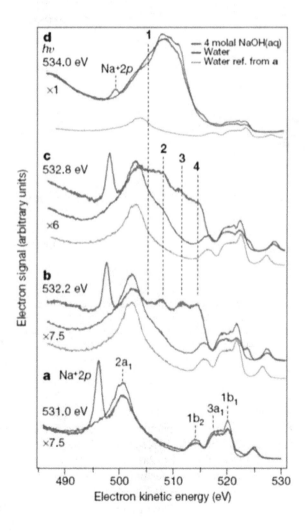

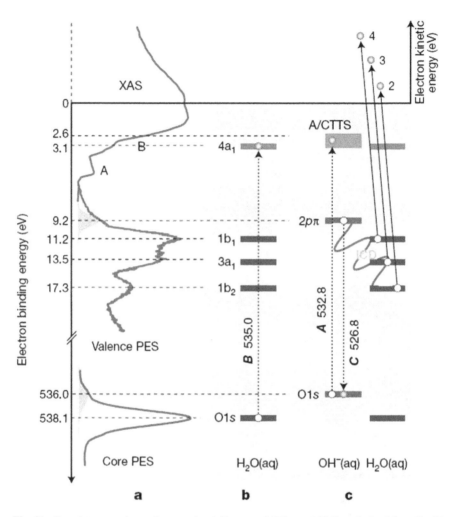

Fig. 21 Complete, experimental energy level diagram of OH_{aq}^- and $H_2O_{aq}^-$ obtained from liquid microjet photoelectron spectroscopy of a 4m NaOH solution. Shown are bound states and, in addition, un-occupied energy levels ($4a_1$ of H_2O and a CTTS of OH_{aq}^-) identified by resonant Auger spectroscopy. Plotted on the left side (**a**) are experimental PES spectra used for the energy level determination; shaded Gaussian peaks (light grey) are the deconvoluted contribution of OH_{aq}^- states (figure reproduced from Ref. [20], Fig. 2)

neat water (blue line) at the 4 photon energies 531.0, 532.2, 532.8, and 534.0 eV. Figure 20a, at 531.0 eV photon energy, shows a reference spectrum taken at an energy slightly below the onset of resonances. In the (blue line) neat water valence spectrum the well known four peaks $2a_1$, $1b_2$, $3a_1$, $1b_1$ are observed and the (red line) $NaOH_{aq}$ spectrum is showing water plus the additional Na_{aq}^+ 2p lines as well as the small valence OH_{aq}^- peak with $E_{kin} = 521.8$ eV. At the subsequent photon energy values at 532.2 eV and at the close by value 532.8 eV the NaOH brine spectra (red) have changed dramatically from the, respective, neat water photoelectron spectra (blue), showing O1s resonance spectator-Auger features in a remarkably high intensity and with very broad additional structure. Further scans for a slightly higher photon energy 534.0 eV, in Fig. 20d, show this is off resonance, already. Except for the OH^- peak and the Na^+, there are no big differences left between the recorded neat water and the NaOH spectrum. A broad and intense peak structure beginning to appear in, both, neat water and in NaOH solution in Fig. 20d arises from a superposition of the water valence state $2a_1$ photoelectron emission peak and an onset of strong water Auger peak structure resulting from excitation of the $4a_1$ unoccupied state of H_2O_{aq} from O1s with excitation centerline energy 535.0 eV. It is designated by the label 1 on top of the spectrum. Within the spectacular OH^- spectator resonance spectra, Fig. 20b, c, three adjacent peak features are recognized with spacings closely matching the energy differences and overall structure between the $1b_2$, $1a_1$ and $1b_1$ valence photoelectron spectra of H_2O. They are marked by the letters 2, 3 and 4 and three additional vertical lines, for underlining the fact that the kinetic energies are constant for Auger electron emission. The apparent, unexpected mixing of valence states of adjacent H_2O molecules into the Auger decay process of core hole excited OH^- is an obviously new phenomenon, difficult to understand [20]. Although it is well known, that the high electrical mobility of the OH^- ion in liquid water has to be attributed to a charge migration process, rather than to molecular diffusion, the time scale is 10^{-12} s for the established charge migration model where a local $OH^- (H_2O)_n$ hydration cluster is rearranging bonds and hopping the charge to a newly formed OH^- center. This mechanism, therefore, must be disputed as being very unlikely because the charge migration step takes orders of magnitudes longer than the decay time of $\sim$ 7 fs, available for the here observed O1s Auger processes. A different, more recent theory attributes the mixing of adjacent molecules' molecular states in an Auger process to a very rapid Intermolecular Coulombic Decay (ICD) phenomenon, with quantitative details for OH^-_{aq} resonance here also still open to discussion.

With the tentative assignments of the OH^-_{aq} resonance emission peak structures number 2, 3 and 4 to Auger emission of water valence electrons $1b_1$, $2a_1$, $1b_2$, in Fig. 20b, c, the picture emerges that one electron from the OH^-_{aq} valence state $2p\pi$ is filling the O1s hole in OH^-_{aq} and, simultaneously, in undisclosed dynamical detail, one of the neighbor-H_2O valence electrons is emitted and observed in the recorded kinetic energy electron spectra. This Auger process interpretation is symbolically depicted in the energy level diagrams at the right-hand side in Fig. 21c. In this energy level scheme are compiled all experimentally determined energies for OH^-_{aq} and, also, all levels for liquid H_2O (in Fig. 21b) which have been obtained from the here described photoelectron spectroscopy measurements. On the left-hand side of the level diagram, Fig. 21a, characteristic photoelectron spectra traces for core and valence states of water and of OH^- in solution are displayed for illustrating the relationship between measured PES features and the herewith determined electron orbital energies in aqueous solution. The also shown energetic positions of the unoccupied excited state $4a_1$ of water in aqueous solution and the new CTTS state for solute OH$^-$ ions are determined with the resonant Auger spectroscopy method, just discussed.

For comparison with more traditional and longer established XAS methods, in the upper part of Fig. 21a is shown, furthermore, a measurement of electronic structure in liquids by X-ray absorption spectroscopy on the K-edge, in the region of O1s excitation to the ionization vacuum level of an aqueous NaOH solution. In the X-ray near edge absorption scan with high resolution tunable x-ray radiation, here the total absorption increases in a large step at the K-edge of water. In addition, pre-edge absorption XAS structures appear at slightly lower photon energies, assigned with the letters A and B and originating from resonance absorptions from unoccupied bound states. The pre-edge structure peak B is related to the excitation of the $4a_1$ unoccupied state of water at 3.1 eV electron binding energy according to the photoelectron energy absolute calibration in Fig. 21b. The peak A observed for OH$^-$ in the XAS spectrum appears at lower excitation energy and its position can be interpreted correctly, only, after it is known from PES results in Fig. 21c that the vertical ionization energy of O1s electrons of the OH_{aq}^- with 536.0 eV BE is smaller by 2.1 eV than the liquid water O1s ionization energy value. Not visible at all in XAS, of course, is the Auger decay transfer from OH$^-$ to vicinal H_2O molecules, following excitation of CTTS state.

7 Concluding Remarks

In going through this retrospective on the development of liquid water microjets for molecular beam studies of evaporating nascent molecular velocity distributions, and then extended, for X-ray photoelectron spectroscopy as diverse as simple determination of valence energies in electrolytes, concentration measurements of surface versus bulk abundance, diagnostics of pH sensitivity of protonation-deprotonation in 1s K shell states individual molecular group atoms, and the detailed spectroscopy of unoccupied near vacuum level states of solvent and solute molecules by resonant Auger spectroscopy, in summary, I feel strongly compelled to thank very many coauthoring colleagues who were with me on this journey at different times over more than three decades. For names I can refer here, only, to the shorter list of coauthors given in the cited references, although it was many more people who have lent their hands, discussed ideas and kept the projects going by their support. Also, I gratefully acknowledge the continuing support by my home institution, Max-Planck-Institut für Strömungsforschung/MPI Dynamics and Self-Organization/, by the Deutsche Forschungsgemeinschaft, by the BESSY synchrotron radiation facility, and by the Max-Born-Institut.

References

1. Gasparus Scotus "Technica Curiosa" p. 172, experimentum XXXVIII; Nuremberg 1664 (Digital Edition: Herzog August Library, Wolfenbüttel)
2. M. Faubel, S. Schlemmer, J.P. Toennies, A molecular beam study of the evaporation of water from a liquid jet. Z Phys D Atoms Molecul Clust **10**, 269–277 (1988)

3. M. Faubel, Photoelectron spectroscopy at liquid surfaces, Chap. 12, vol. I, in *Photoionization and photodetachment*, vol. 101A, ed. by C.Y. Ng (World Scientific Publishing, Singapore, 2000), pp. 634–690

4. S. Hess, M. Faubel "Gase und Molekularstrahlen" Fig. 1.70, in: *Bergmann-Schäfer Lehrbuch der Experimentalphysik* Bd. 5, ed. by K. Kleinermanns, Walter de Gruyter (Berlin, New York 2006), p. 119

5. O. Stern, Nachtrag zu meiner Arbeit: Eine direkte Messung der thermischen Molekularstrahlgeschwindigkeit (Comment to my paper: A direct measurement of the thermal molecular beams velocity). Z. f. Phys. **3**, 417–421 (1920) "Zusammenfassung: Es wird die Frage der Geschwindigkeitsverteilung der von einer Flüssigkeitsoberfläche ausgehenden Moleküle diskutiert. Ferner werden die Resultate einiger neuerer Messungen der mittleren Geschwindigkeit von Silberatomen mitgeteilt" 'Summary: I discuss the question of the velocity distribution of molecules emerging from the surface of a liquid. In addition, results of some new measurements of the mean velocity of silver atoms are communicated."

6. M. Faubel, T. Kisters, Non-equilibrium molecular evaporation of carboxylic acid dimers. Nature **339**, 527–529 (1989)

7. D. Chandler, *Introduction to Modern Statistical Mechanics* (Oxford University Press, New York, 1987)

8. M.P. Allen, D.J. Tildesley, *Computer Simulation of Liquids* (Oxford University Press, Oxford, 1987)

9. J.R. Rumble (ed.), *CRC handbook of chemistry and physics*, 98th edn. (CRC Press, Taylor & Francis Ltd, Boca Raton, London, 2017)

10. M. Faubel, B. Steiner, J.P. Toennies, Photoelectron spectroscopy of liquid water, some alcohols, and pure nonane in free micro jets. J. Chem. Phys. **106**, 9013 (1997)

11. M. Faubel, B. Winter, Photoemission from liquid aqueous solutions. Chem. Rev. **106**(4), 1176–1211 (2006)

12. W.S.M. Werner, W. Smekal, C.J. Powell, *NIST database for simulation of electron spectra for surface analysis* (U.S. Department of Commerce, National Institute of Standards and Technology, Gaithersburgh MD, 2005)

13. R. Weber, B. Winter, P.M. Schmidt, W. Widdra, I.V. Hertel, M. Dittmar, M. Faubel, Photoemission from aqueous alkali-metal—iodide salt solutions using EUV synchrotron radiation. J. Phys. Chem. B **108**(15), 4729–4736 (2004)

14. N. Ottosson, M. Faubel, S.E. Bradforth, P. Jungwirth, B. Winter, Photoelectron spectroscopy of liquid water and aqueous solution: Electron effective attenuation length and emission-angle anisotropy. J. El. Rel. Phen. **177**, 60–70 (2010)

15. P. Jungwirth, D.J. Tobias, The molecular structure of salt solutions. J. Phys. Chem. B **105**, 10468–10472 (2001)

16. D. Nolting, N. Ottosson, M. Faubel, I.V. Hertel, B. Winter, Pseudoequivalent nitrogen atoms in aqueous imidazole distinguished by chemical shifts in photoelectron spectroscopy. J. Am. Chem. Soc. **130**(26), 8150–8151 (2008)

17. T. Lewis, M. Faubel, B. Winter, J.C. Hemminger, CO_2 Capture in amine-based aqueous solution: role of the gas–solution interface. Angew. Chem. Int. Ed. **50**, 10178 (2011)

18. B. Winter, M. Faubel, I.V. Hertel, C. Pettenkofer, S.E. Bradforth, B. Jagoda-Cwiklik, L. Cwiklic, P. Jungwirth, Electron binding energies of hydrated H_3O+ and $OH-$: photoelectron spectroscopy of aqueous acid and base solutions combined with electronic structure calculations. J. Am. Chem. Soc. **128**(12), 3864–3865 (2005)

19. B. Winter, E.F. Aziz, N. Ottosson, M. Faubel, N. Kosugi, I.V. Hertel, Electron dynamics in charge-transfer-to-solvent states of aqueous chloride revealed by Cl^- 2p resonant auger-electron spectroscopy. J. Am. Chem. Soc. **130**(22), 7130–7138 (2008)

20. E.F. Aziz, N. Ottosson, M. Faubel, I.V. Hertel, B. Winter, Interaction between liquid water and hydroxide revealed by core-hole de-excitation. Nature **455**, 89–91 (2008)

6

Stern-Gerlach Interferometry with the Atom Chip

**Mark Keil, Shimon Machluf, Yair Margalit, Zhifan Zhou, Omer Amit,
Or Dobkowski, Yonathan Japha, Samuel Moukouri, Daniel Rohrlich,
Zina Binstock, Yaniv Bar-Haim, Menachem Givon, David Groswasser,
Yigal Meir and Ron Folman**

Abstract In this invited review in honor of 100 years since the Stern-Gerlach (SG) experiments, we describe a decade of SG interferometry on the atom chip. The SG effect has been a paradigm of quantum mechanics throughout the last century, but there has been surprisingly little evidence that the original scheme, with freely propagating atoms exposed to gradients from macroscopic magnets, is a fully coherent quantum process. Specifically, no full-loop SG interferometer (SGI) has been realized with the scheme as envisioned decades ago. Furthermore, several theoretical studies have explained why it is a formidable challenge. Here we provide a review of our SG experiments over the last decade. We describe several novel configurations such as that giving rise to the first SG spatial interference fringes, and the first full-loop SGI realization. These devices are based on highly accurate magnetic fields, originating from an atom chip, that ensure coherent operation within strict constraints described by previous theoretical analyses. Achieving this high level of control over magnetic gradients is expected to facilitate technological applications such as probing of surfaces and currents, as well as metrology. Fundamental applications include the probing of the foundations of quantum theory, gravity, and the interface of quantum mechanics and gravity. We end with an outlook describing possible future experiments.

M. Keil · S. Machluf · Y. Margalit · Z. Zhou · O. Amit · O. Dobkowski · Y. Japha · S. Moukouri ·
D. Rohrlich · Z. Binstock · Y. Bar-Haim · M. Givon · D. Groswasser · Y. Meir · R. Folman (✉)
Department of Physics, Ben-Gurion University of the Negev, Be'er Sheva 84105, Israel
e-mail: folman@bgu.ac.il

M. Keil
e-mail: mhkeil@gmail.com

S. Machluf
Analytics Lab, Amsterdam, The Netherlands

Y. Margalit
Research Laboratory of Electronics, MIT-Harvard Center for Ultracold Atoms, Department
of Physics, Massachusetts Institute of Technology, Cambridge, MA 02139, USA

Z. Zhou
Joint Quantum Institute, National Institute of Standards and Technology and the University
of Maryland, College Park, Maryland 20742, USA

1 Introduction

This review follows the centennial conference held in Frankfurt in the same building housing the original Stern-Gerlach (SG) experiments. Here we describe the SG interferometry performed in our laboratories at Ben-Gurion University of the Negev (BGU) over the last decade.

The trail-blazing experiments of Otto Stern and Walther Gerlach one hundred years ago [1–4] required a few basic ingredients: a source of isolated atoms with well-specified momentum components (provided by their atomic beam), an inhomogeneous magnetic field and, if we follow the historical account of events in [5], also a smoky cigar. In this review, we present our approach to these first two ingredients, with our sincere apologies that we will not be able to adequately address the third.

As Dudley Herschbach notes [4], the SG experiments formed the basis for a "symbiotic entwining of molecular beams with quantum theory" and, as shown in many of the papers at this centennial conference, this symbiotic relationship remains vigorous to the present day. In this review, our source of isolated atoms is instead provided by the new world of ultra-cold atomic physics, to which we couple inhomogeneous magnetic fields that are provided naturally by an atom chip [6]. Current-carrying wires on such chips were first realized as magnetic traps for ultra-cold atoms at the turn of the (twenty-first) century [7–9] and reviewed extensively since [6, 10–14]. We are using the atom chip as our basis for coherently manipulating atoms in a way that is complementary to the atomic and molecular beam techniques pioneered by Otto Stern and practiced so energetically and creatively by his scientific descendants.

The work presented here is performed with high-quality atom chips fabricated by our nano-fabrication facility [15]. The atom chip is advantageous for Stern-Gerlach interferometry (SGI) for 4 main reasons. First, the source (Bose-Einstein condensates, BEC) is a minimal-uncertainty wavepacket so it is very well defined in position and momentum. Second, the source of the magnetic gradients (current-carrying wires on the atom chip) is very well aligned relative to the atomic source. Third, due to the very small atom-chip distance, the gradients are very strong, and significant Stern-Gerlach splitting can be realized in very short times. Fourth, the gradients are very well defined in time since there are no coils and the inductance of the chip wires is negligible. We will describe how these advantages have overcome long-standing difficulties and have enabled different SG configurations to be realized at BGU (e.g., spatial interference patterns [16, 17] and a "full-loop" SGI [18, 19]) alongside several applications, such as spatially splitting a clock [20, 21]. Finally, let us mention that while the interferometers presented here are of a new type, it is worthwhile noting decades of progress in matter-wave interferometry [22].

The discovery of the Stern-Gerlach (SG) effect [1] was followed by ideas concerning a full-loop SGI that would consist of freely propagating atoms exposed to magnetic gradients from macroscopic magnets. However, starting with Heisenberg [23], Bohm [24] and Wigner [25] considered a coherent SGI impractical because it was thought that the macroscopic device could not be made accurate enough to ensure a reversible splitting process [26]. Bohm, for example, noted that the magnet would need to have "fantastic" accuracy [24]. Englert, Schwinger and Scully analyzed the problem in more detail and coined it the Humpty-Dumpty[1] (HD) effect [28–31]. They too concluded that for significant coherence to be observed, exceptional accuracy in controlling magnetic fields would be required. Indeed, while atom interferometers based on light beam-splitters enjoy the quantum accuracy of the photon momentum transfer, the SGI magnets not only have no such quantum discreteness, but they also suffer from inherent lack of flatness due to Maxwell's equations [32]. Later work added the effect of dissipation and suggested that only low-temperature magnetic field sources would enable an operational SGI [33]. Claims have even been made that no coherent splitting is possible at all [34].

Undeterred, we utilize the novel capabilities of the atom chip to address these significant hurdles. Let us briefly preview our most recent and most challenging realization, the full-loop SGI, in which magnetic field gradients act on the atom during its flight through the interferometer, first splitting, and then re-combining, the atomic wavepacket. We obtain a high full-loop SGI visibility of 95% with a spin interference signal [18, 19] by utilizing the highly accurate magnetic fields of an atom chip [6]. Notwithstanding the impressive endeavors of [35–45] this is, to the best of our knowledge, the first realization of a complete SG interferometer analogous to that originally envisioned a century ago.

Achieving this high level of control over magnetic gradients may facilitate fundamental research. Stern-Gerlach interferometry with mesoscopic objects has been suggested as a compact detector for space-time metric and curvature [46], possibly enabling detection of gravitational waves. It has also been suggested as a probe for the quantum nature of gravity [47]. Such SG capabilities may also enable searches for exotic effects like the fifth force or the hypothesized self-gravitation interaction [48]. We note that the realization presented here has already enabled the construction of a unique matter-wave interferometer whose phase scales with the cube of the time the atom spends in the interferometer [19], a configuration that has been suggested as an experimental test for Einstein's equivalence principle when extended to the quantum domain [49].

High magnetic stability and accuracy may also benefit technological applications such as large-momentum-transfer beam splitting for metrology with atom interferometry [50–52], sensitive probing of electron transport, e.g., squeezed currents [53], as well as nuclear magnetic resonance and compact accelerators [54]. We note that since the SGI makes no use of light, it may serve as a high-precision surface probe at short distances for which administering light is difficult.

For the purpose of this review, it is especially important to also realize that the atom chip allows our atoms to be completely isolated from their environment. This

---

[1]Can a fragile item be taken apart and be re-assembled perfectly? ... another tough problem, according to the popular English rhyme [27]
Humpty Dumpty sat on a wall,
Humpty Dumpty had a great fall.
All the king's horses
And all the king's men
Couldn't put Humpty together again.

is demonstrated, for example, by the relatively long-term maintenance of spatial coherence that can be achieved despite a temperature gradient from 300 K to 100 nK over a distance of just 5 μm [55]. Coherence of internal degrees of freedom close to the surface has also been measured to be very high [56].

This review is organized into the following sections:

Section 2. Particle Sources: a brief discussion of how the atom chip complements and extends the century-long use of atomic and molecular beams in Stern-Gerlach experiments;

Section 3. The Atom Chip Stern-Gerlach Beam Splitter: detailing relevant aspects of the atom chip and its basic operating characteristics as a platform for SGI;

Section 4. Half-Loop Stern-Gerlach Interferometer: first realization of SGI with spatial fringe patterns;

Section 5. Full-Loop Stern-Gerlach Interferometer: first realization of the four-field complete SGI with spin population fringes;

Section 6. Applications: clock interferometry and complementarity, the matter-wave geodesic rule and geometric phase, and a T^3 interferometer realizing the Kennard phase;

Section 7. Outlook: extending the atom-chip based SGI experiments to ion beams and to massive particles.

Finally, we note that the SG effect, in conjunction with the atom chip, may also be used for novel applications without the use of interferometry. For example, we have used the SG spin-momentum entanglement to realize a novel quantum work meter. In this work, done in conjunction with the group of Juan Pablo Paz, we were able to test non-equilibrium fluctuation theorems [57].

As we hope to show in this review, we believe that the atom chip provides a novel and powerful tool for SG interferometry, with much yet to learn as SG studies enter their second century. May we continue to find surprises, fundamental insights, and exciting applications.

2 Particle Sources

Molecular beam experiments exhibiting quantum interference, diffraction, and reflection have been brought very skillfully into the modern era in presentations at this Conference by Markus Arndt, Maksim Kunitski, and Wieland Schöllkopf, and as outlined in the keynote address by Peter Toennies. In particular, Stern's vision—and realization—of diffraction of atomic and molecular beams (see, for example [4]) have found their modern expression in the work of all these experts, and many others. Here we will concentrate on a complementary approach to precisely specify internal and external quantum states and how they can be used to study interference phenomena in particular.

Table 14.1 Parameters relating to diffraction experiments using He atomic beams [58, 59], Talbot-Lau interference experiments with macromolecules [60], and interference experiments using BEC's [17] as described in this review

Type	Source	Species	Temperature (K)	σ_z (μm)	σ_{v_z} (mm/s)	$k = \sigma_{p_z}/\hbar$ (μm^{-1})	$\sigma_z \sigma_{p_z}/\hbar$	Ref.
Diffraction	Beam	^{4}He	10^{-3}	20	14	0.9	18	[58]
Diffraction	Beam	^{4}He	Not given	50	43	2.6	130	[59]
T-L interference[a]	Beam	Macromolecules	Not given	0.266	0.04	16	4.3	[60]
Interference	BEC	^{87}Rb	40×10^{-9}	6	2.8	3.8	23	[17]
Particle-on-demand[b]	Ion trap	^{40}Ca$^+$	–	0.006	900	5×10^6	3×10^4	[61, 62]
First realization	Beam	Ag	1300	30	230	400	1×10^4	[1]

The temperature shown for the beam experiments corresponds to the velocity spread superimposed on the moving frame of the longitudinal most-probable velocity. The position spread σ_z for the He beams is the beam collimator width, while the velocity spread σ_{v_z} is calculated from the beam angular divergence and its most-probable longitudinal velocity [63] ($v_x = v_{mp} \approx 288$ m/s for a He beam source temperature of 8 K). For the macromolecular beam, the parameters are taken from the grating period, interferometer length, and the stated longitudinal deBroglie wavelength. Corresponding parameters for the BEC are calculated using the Thomas-Fermi approximation and a temperature at which the BEC is about 90% pure. Parameters for the original Stern-Gerlach experiment are shown for comparison in the last line. All species are in their ground electronic state. The x- and z- co-ordinates refer to the horizontal and vertical directions respectively, where the beam experiments are horizontal (so z is the transverse direction) while the BEC experiments are vertical (so z is the longitudinal direction). We do not give parameters for the "beaded atom" experiments [36] since we believe that spatial interference fringes were not observed, as explained in [64]

[a]Talbot-Lau interference, as applied to matter-wave interference studies, is described in detail in [65]. The particle species in the quoted study are functionalized oligoporphyrin macromolecules with up to 2000 atoms and masses > 25000 Da [60].

[b]These parameters are for a proposal for SGI using ion beams that will be discussed in Sect. 7.1.

Let us begin by comparing experimental parameters used in the ultra-cold atomic environment in our laboratory, typically achieved with BECs of ^{87}Rb, with corresponding state-of-the-art parameters for atomic beams. Table 14.1 summarizes parameters that are most relevant for these experiments. Note that the beam experiments are conducted in a horizontal plane, transverse to the beam propagation direction, while our BEC interference experiments are conducted in an exclusively longitudinal direction with the atoms falling vertically due to gravity (and with all applied forces also acting in the longitudinal direction).

We see that ultra-cold atom localization and velocity spreads are on the same order as transverse localization from the exemplary atomic and molecular beam experiments quoted here but, of course, ultra-cold atoms are also localized in all three dimensions, whereas the beam techniques do not achieve localization along the beam propagation axis.

3   The Atom Chip Stern-Gerlach Beam Splitter

In order to apply Stern-Gerlach splitting, our ultra-cold atomic sample needs to have at least two spin states. However, our initial atomic sample is purely in the $|F, m_F\rangle = |2, 2\rangle$ state of ^{87}Rb. After preparing a BEC on the atom chip, our SG implementation therefore begins by first releasing the magnetic trap, and then applying a radio-frequency (RF) $\pi/2$ Rabi pulse to create an equal superposition of the two internal spin states $\frac{1}{\sqrt{2}}(|1\rangle + |2\rangle)$, where $|1\rangle$ and $|2\rangle$ represent the $m_F = 1$ and $m_F = 2$ Zeeman sub-levels of the $F = 2$ manifold in the ground electronic state [66]. Transitions to other m_F levels are avoided by retaining a modest homogeneous magnetic field even after trap release. A field of about 30 G is sufficient to create an effective two-level system by pushing the $m_F = 0$ sub-level about 200 kHz out of resonance with the $|2\rangle \rightarrow |1\rangle$ RF transition due to the non-linear Zeeman effect. The intensity of the RF Rabi pulses is calibrated such that a pulse duration of 20 μs corresponds to a complete population inversion between the two states, i.e., a π-pulse. This corresponds to a Rabi frequency of $\Omega_{RF} = 2\pi \cdot 25$ kHz.

We now consider the second factor crucial to the success of our SGI experiments: fast and precise magnetic fields, in both magnitude and direction, may be delivered by pulsed currents passed through micro-fabricated wires on the atom chip. Simple Biot-Savart considerations for atom chip wires, as used in our experiments, yield magnetic field gradients of about 200 G/mm at $\sim$100 μm from the chip, which is the starting distance for most of our experiments. Accurate control of this initial position, which is also crucial for the success of the experiments, is ensured by accurate control of chip wire currents and the homogeneous magnetic field referred to above. In addition, the straight atom chip wires have very low inductance, thereby enabling the generation of well-defined magnetic force pulses with currents that are typically tens of μs long. Such pulses are, in principle, able to induce momentum changes of hundreds of $\hbar k$.[2] Our earliest implementations of these experimental characteristics [67] were improved in subsequent apparatus upgrades [64].

Since the experiments proceed after turning off the magnetic trap, the observation time is limited by the time-of-flight (TOF) of the falling atoms and the field-of-view of our absorption imaging detection system. The latter is limited to about 4 mm, corresponding to a maximum TOF of about 28 ms. The optical detection system has a spatial resolution of about 5 μm, an important consideration for measuring spatial interference patterns (Sect. 4). Further experimental details may be found in several recent Ph.D. theses from our laboratory [64, 67, 68].

The Stern-Gerlach beam splitter (SGBS), first implemented in [16], begins with an equal superposition of $|1\rangle$ and $|2\rangle$ as described above and depicted schematically in Fig. 1. We then apply a magnetic field gradient $\nabla|\mathbf{B}|$ for duration T_1, which creates a state-dependent force $\mathbf{F}_{m_F} = m_F g_F \mu_B \nabla|\mathbf{B}|$ on the atomic ensemble, where μ_B, g_F, and m_F denote the Bohr magneton, the Landé factor, and the projection of the angular momentum on the quantization axis, respectively.

The magnetic potential created by the atom chip can be approximated as a sum of a linear part with characteristic force $\mathbf{F}$ and a quadratic part with characteristic frequency ω. After this magnetic gradient splitting pulse, the new state of the atoms is given by $\psi_f = \frac{1}{\sqrt{2}}(|1, p_1\rangle + |2, p_2\rangle)$, where $\mathbf{p}_i = \mathbf{F}_i T_1$ ($i = 1, 2$). This state represents a coherent superposition of two distinct momentum states, which are then allowed to separate spatially, thereby completing the operation of momentum and spatial splitting.

As we discuss further in the following sections, the SGBS can be extended as a tool for SGI. We describe two main configurations: a "half-loop" configuration in which the separated wavepackets are allowed to propagate freely, expand and eventually overlap, producing spatial interference patterns analogous to a double-slit experiment, and a "full-loop" configuration in which the wavepackets are actively re-combined, analogous to a Mach-Zehnder interferometer.

By applying additional pulses with different timing, these methods have been used to demonstrate, to the best of our knowledge, the first Stern-Gerlach spatial fringe interferometer (Sect. 4, [16, 17]), the first full-loop Stern-Gerlach interferometer (Sect. 5, [18, 19]), and several applications that we will describe in Sect. 6, including experiments to simulate the effect of proper time on quantum clock interference [20, 21].

4 Half-Loop Stern-Gerlach Interferometer

The two separated wavepackets generated by the SGBS initiate the pulse sequence shown in Fig. 2. Just after the SG splitting pulse, another RF $\pi/2$ pulse (10 μs dura-

---

[2] We express the momentum transfer in units of $\hbar k$, a reference momentum of a photon with 1 μm wavelength, in order to compare with atom interferometry based on optical beam splitters.

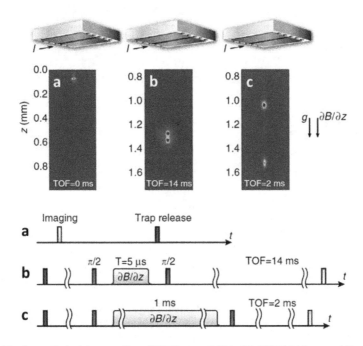

Fig. 1 The Stern-Gerlach beam-splitter (SGBS) at work [16, 67]. SGBS (a) input and (b, c) output images, and the corresponding schematic descriptions. The top row depicts our atom chip, with a pulsed current I being used to generate the magnetic gradient $\partial B/\partial z$ (we currently use three parallel wires with equal currents but opposing polarities). The chip faces downwards so that atoms can separate vertically during their free fall. (a) A magnetically trapped BEC in state $|2\rangle$ before release. (b) After a weak splitting of less than $\hbar k$ using a $5\,\mu$s magnetic gradient pulse and allowing a TOF of 14 ms. (c) After a strong splitting of more than $40\,\hbar k$ using a 1 ms magnetic gradient pulse and allowing a TOF of 2 ms. Interferometric signals are formed either as spatial interference fringes by passively allowing overlap of the wavepackets (the "half-loop" SGI), or as spin-state population oscillations upon actively recombining them (the "full-loop" SGI), as described in Sects. 4 and 5 respectively. Adapted from [16].

tion) is applied, creating a wavefunction consisting of four wavepackets [67], of which we are concerned only with the two $|2\rangle$ wavepackets having momenta $\mathbf{p}_1$ and $\mathbf{p}_2$ (the $|1\rangle$ components can be disregarded since they appear at different final positions on completing the pulse sequence and a TOF).

The time interval between the two RF pulses (in which there are only two wavepackets, each having a different spin) is reduced to a minimum ($\sim 40\mu s$) to suppress the hindering effects of a noisy and uncontrolled magnetic environment, thereby removing the need for magnetic shielding. After a magnetic gradient pulse of duration T_2, designed to stop the relative motion of the two wavepackets, the atoms fall under gravity for a relatively long TOF, expanding freely until they overlap to create spatial interference fringes as shown schematically in Fig. 2 and experimentally in Fig. 3.

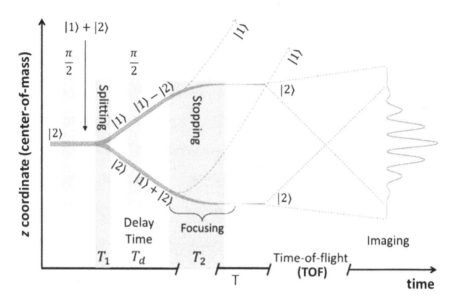

Fig. 2 Schematic depiction of the longitudinal half-loop SGI giving rise to spatial interference fringes (vertical position z in the center-of-mass frame vs. time). The initial wavepacket $|2\rangle$ (extreme left) is subjected to a $\pi/2$ pulse (blue column) that transfers the atoms into the superposition state $|1\rangle + |2\rangle$. The first magnetic gradient pulse of duration T_1 (purple column) induces a Stern-Gerlach splitting into $|1\rangle$ (green curve) and $|2\rangle$ (purple curve) having momenta $\mathbf{p}_1$ and $\mathbf{p}_2$, respectively. We then immediately apply a second $\pi/2$ pulse that places these diverging $|1\rangle$ and $|2\rangle$ states into equal superpositions $|1\rangle \mp |2\rangle$ as shown. The delay time T_d allows these wavepackets to spatially separate (in the z direction). The duration T_2 of a second gradient pulse is tuned to bring the momentum difference between the $|2\rangle$ components close to zero (see text), allowing their space-time trajectories to become parallel (solid purple curves) while expelling the $|1\rangle$ components (dotted green trajectories). The atoms then fall freely under gravity. Given sufficient time-of-flight, the two $|2\rangle$ wavepackets expand (dotted purple lines) and eventually overlap to generate spatial interference fringes, which are measured by taking an absorption image of the atoms. We note that due to the curvature of the magnetic field forming the magnetic gradient pulse, the long T_2 pulse also focuses the wavepackets, as depicted in the figure. In fact, this focusing accelerates the process of final expansion, thereby creating the two-wavepacket overlap in a shorter time. Adapted from [17] with permission © IOP Publishing & Deutsche Physikalische Gesellschaft. CC BY 3.0

The period of the interference fringes must be large enough to be observable with the spatial resolution of our imaging system (about 5 μm). This is accomplished if two conditions are fulfilled. First, the distance between the two wavepackets, d, should not be too large, since in principle the fringe periodicity varies as ht/md when the relative momentum is zero, where h, t, and m are the Planck constant, TOF duration, and the atomic mass, respectively. Second, the momentum difference between the two wavepackets should be smaller than their momentum width to avoid orthogonality. This is accomplished by tuning the duration T_2 of the second gradient pulse, which can stop the relative motion of the two $|2\rangle$ wavepackets; despite being in the same spin state, the slower wavepacket experiences a stronger impulse than

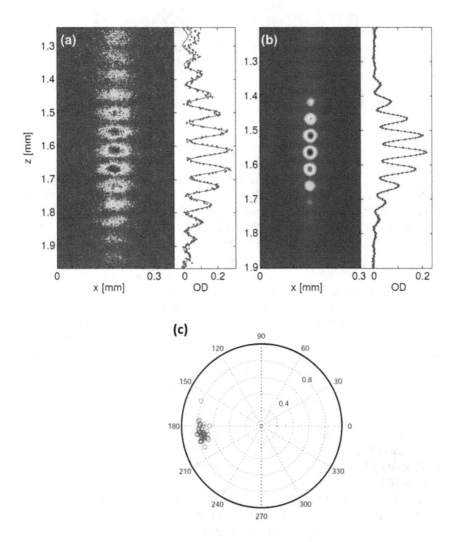

Fig. 3 Spatial interference patterns from the Stern-Gerlach interferometer. **a** A single-shot interference pattern of a thermal cloud with a negligible BEC fraction, fitted to Eq. (1) with a visibility of $V = 0.65$ (only slightly lower than single-shot visibilities typically measured for a BEC). **b** A multi-shot image made by averaging 40 consecutive interference images using a BEC (no correction or post-selection) with a normalized visibility of $V_N = 0.99$. **c** Polar plot of phase $0° \leq \phi \leq 360°$ versus visibility $0 \leq V \leq 1$ obtained from fitting each of the 40 consecutive images averaged in (**b**). The experimental parameters are $(T_1, T_d, T_2) = (4, 116, 200)\ \mu$s. Adapted from [64]

the faster one since it is considerably closer to the atom chip after the relatively long delay time T_d. We have found that zeroing the momentum difference between the two wavepackets is very robust [67].

Given that the final momentum difference between the two interfering wavepackets is smaller than their momentum spread, they overlap after a sufficiently long TOF and an interference pattern appears with the approximate form:

$$n(z, t) = A \exp \left[-\frac{(z - z_{CM})^2}{2\sigma_z(t)^2} \right]$$
$$\times \left[1 + V \cos \left(\frac{2\pi}{\lambda}(z - z_{ref}) + \phi \right) \right], \qquad (1)$$

where A is the amplitude, z_{CM} is the center-of-mass (CM) position of the combined wavepacket at the time of imaging, $\sigma_z(t) \approx \hbar t/2m\sigma_0$ is the final Gaussian width, $\lambda \approx 2\pi \hbar t/md$ is the fringe periodicity ($d = |z_1 - z_2|$ is the distance between the wavepacket centers), V is the interference fringe visibility, and $\phi = \phi_2 - \phi_1$ is the global phase difference. The vertical position z is relative to a fixed reference point z_{ref}. The phases ϕ_1 and ϕ_2 are determined by an integral over the trajectories of the two wavepacket centers. We emphasize that Eq. (1) is not a phenomenological equation, but rather an outcome of our analytical model [16].

In order to characterize the stability of the phase, which is the main figure of merit in interferometry, we average multiple experimental images with no postselection or alignment (each single-shot image is a result of one experimental cycle). Large fluctuations in the phase and/or fringe periodicity in a set of single-shot images would result in a low multi-shot visibility, while small fluctuations correspond to high multi-shot visibility. The multi-shot visibility is therefore a measure of the stability of the phase and periodicity. Single-shot and multi-shot visibilities are all extracted by fitting to Eq. (1) after averaging the experimental images along the x direction (see Fig. 3) to reduce noise. We note that these procedures have been used over several years of half-loop SGI studies [16, 17], while the experimental results were simultaneously being greatly improved by significant modifications to the original apparatus [64, 67].

For a pure superposition state, as in our model, perfect fringe visibility V would be 1. A quantitative analysis of effects reducing V appears in [17, 64]. Some of these effects are purely technical, e.g., imperfect BEC purity and wavepacket overlap in 3D, as well as various imaging limitations etc. Such technical effects are irrelevant to the phase and periodicity stability shown by the multi-shot visibility, so we normalize the latter to the mean of the single-shot visibilities taken from the same sample: $V_N \equiv V_{avg}/\langle V_s \rangle$, where V_{avg} is the (un-normalized) visibility of the multi-shot average extracted from the fit and $\langle V_s \rangle$ is the mean visibility of the single-shot images which compose that multi-shot image. The normalized multi-shot visibility thus reflects shot-to-shot fluctuations of the global phase ϕ and the fringe periodicity λ. We note that some BEC intrinsic effects, such as phase diffusion, would not lead to a reduction of the single-shot visibility, but may cause the randomization of the shot-to-shot phase. However, such effects are expected to be quite weak, since atom-atom interactions rapidly become negligible as the BEC expands in free-fall, and the experiment may be described by single-atom physics.

Representative results from the above analysis are shown in Fig. 3. The very high (normalized) visibility shown in (b) demonstrates that the phase and periodicity are highly reproducible for each experimental cycle, the former being particularly emphasized in plot (c). High-visibility fringes ($V > 0.90$) were observed over a wide variety of experimental parameters, covering a range of maximum separations and velocities between the wavepackets. In particular, we conducted experiments at the apparatus-limited maximum value of $T_d = 600\,\mu s$ (which also required a long TOF $= 21.45$ ms) in order to maximize the spatial separation of the wavepackets during their time in the interferometer. These measurements achieved a separation $d = 3.93\,\mu m$, a factor of 20 larger than the atomic wavepacket size (after focusing, see Fig. 2), while maintaining a normalized visibility of $V_N = 0.90$ [17].

Given that our observed stable interference fringes arise from such well-separated paths, these experiments demonstrate what is, to the best of our knowledge, the first implementation of spatial SG interferometry. This achievement is due to three main differences compared with previous SG schemes. Firstly, we have used minimal-uncertainty wavepackets (a BEC) rather than thermal beams. Secondly, while the splitting is based on two spin states, the wavepackets in the two interferometer arms are in the same spin state for most of the interferometric cycle, thus reducing their sensitivity to disruptive external magnetic fields. Finally, chip-scale temporal and spatial control allows the cancellation of path difference fluctuations. It should also be noted that a longitudinal SGI, based on a particle beam source, cannot take images of spatial fringes due to the high velocity of the fringe pattern in the lab frame.

This, however, is not yet the four-field SGI originally envisioned shortly after the original Stern-Gerlach experiments (as recounted in [26]), since the separated wavepackets are not actively recombined in both position and momentum. The two remaining magnetic gradients required to complete such a "closed" SGI are discussed in the following section.

5 Full-Loop Stern-Gerlach Interferometer

Clearly, if a wavepacket can be coherently reconstructed after SG splitting and recombination in a four-field configuration [26], it should be possible to observe an interference pattern at the output of such an SGI. To the best of our knowledge however, no such interference pattern has heretofore been measured experimentally, and this is the task that we now describe, many details of which are taken from [64] and references therein.

The device envisioned consists of four successive regions of magnetic gradients giving rise to the operations of splitting, stopping, reversing and, finally, stopping the two wavepackets, as shown schematically in Fig. 4a. If executed perfectly, the two wavepackets would arrive at the output of such an interferometer with a minimal relative spatial displacement and momentum difference, so that an arbitrary initial spin state should be recoverable, using the spin state of the recombined wavepacket as the interference signal. However, the operation of such an interferometer was

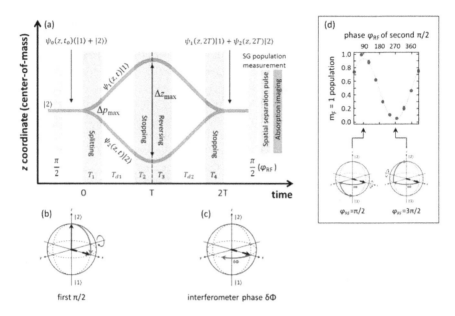

Fig. 4 The longitudinal full-loop SGI giving rise to spin population oscillations, plotted in the center-of-mass frame as in Fig. 2. **a** The sequence consists of RF pulses (blue) to manipulate the inner (spin) degrees of freedom and magnetic gradients (purple) to control the momentum and position of the wavepackets. The interferometer is prepared from the initial wavepacket $|2\rangle$ (extreme left) by applying a $\pi/2$ pulse that transfers the atoms into the superposition state $|1\rangle + |2\rangle$ [Bloch sphere shown in (**b**)]. The first magnetic gradient pulse at $t = 0$ induces a Stern-Gerlach splitting into $|1\rangle$ (green curve) and $|2\rangle$ (purple curve). Three additional magnetic gradient pulses are used to stop the relative motion of the wavepackets (at their maximum separation Δz_{max}), reverse their momenta, and finally stop them at the same position along z. The re-combined wavepacket at $t = 2T$ is therefore written as $\psi_1(z, 2T)|1\rangle + \psi_2(z, 2T)|2\rangle$, shown in (**c**) for an arbitrary interferometer phase $\delta\Phi$. After recombination, the population in $|1\rangle$ is measured by applying a second $\pi/2$ pulse with variable phase φ_{RF}, followed by a magnetic gradient to separate the populations and a subsequent pulse of the imaging laser. We expect to observe spin population fringes, i.e., oscillations in the $m_F = 1$ population, as we scan φ_{RF}, as indeed shown by the experimental results in (**d**), for which the measured visibility is 95%. The Bloch spheres in (**d**) show the particular case in which the initial vector (dashed black arrow) acquires an interferometer phase $\delta\Phi = \pi/2$ (blue arrow) followed by rotations about the $+x$ ($\varphi_{RF} = \pi/2$) or $-x$ ($\varphi_{RF} = 3\pi/2$) axes respectively (red arrows). The states $|F, m_F\rangle = |2, 2\rangle \equiv |2\rangle$ and $|2, 1\rangle \equiv |1\rangle$ are defined along the z axis in the Bloch spheres. Adapted from [64].

considered to be technically impractical, since coherent recombination of the two beam paths would require extremely precise control of the magnetic fields [24].

Our experiments begin, as before, with a $\pi/2$ pulse creating a superposition of the two spin states $|1\rangle$ and $|2\rangle$ of ^{87}Rb that is subsequently split into two momentum components by a magnetic gradient pulse (along the vertical axis z) as described in Sects. 3 and 4. Additional magnetic gradient pulses are needed to "close" the loop of such an interferometer, i.e., to overlap the wavepackets spatially and with zero relative momentum. To stop the relative motion of the two wavepackets after the first pulse, and to accelerate them backwards, we reverse the current on the atom chip, causing the force applied by the magnetic field gradient to be in the opposite direction. Alternatively, we can apply a spin inversion procedure by using a π Rabi pulse that inverts the population between the two internal states, following which a magnetic gradient pulse will then apply the opposite differential momentum to the two wavepackets. We obtain the signal with the help of a second $\pi/2$ pulse, followed by a spin population measurement. We measure the visibility by scanning the phase φ_{RF} of this $\pi/2$ pulse.

Our full-loop interferometer is implemented with an experimental system in which care is taken to reduce a wide range of hindering effects relative to our earliest work [16]. For example, a new atom chip was installed, utilizing a 3-wire configuration to produce a quadrupole magnetic field whose zero is at the precise height of the BEC. This reduces phase fluctuations by exposing the wavepackets to a weaker magnetic field while still generating strong magnetic gradients.

The practical difficulty encountered in re-assembling the original wavefunction was named the Humpty-Dumpty (HD) effect [28–30], implying that the initial wavepacket breaks under the SG field and cannot be reunited, as noted in the brief historical perspective given in Sect. 1. Quantitatively, the spin coherence, which is measurable as the visibility V of the observed spin fringes, is expressed as [29]

$$V = \exp\left\{-\frac{1}{2}\left[\left(\frac{\Delta z(2T)}{\sigma_z}\right)^2 + \left(\frac{\Delta p_z(2T)}{\sigma_p}\right)^2\right]\right\}, \qquad (2)$$

where $\Delta z(2T)$ and $\Delta p_z(2T)$ denote the mismatch between the wavepackets in their final position and momentum respectively, after the interferometer duration $2T$ (Fig. 4a), and σ_z and σ_p are the corresponding initial wavepacket widths. Equation (2) summarizes the main result of the HD papers in relation to our experimental observable. We emphasize that this reduction in visibility has nothing to do with effects of decoherence due to some coupling with the environment. We also note that the above HD calculation is done for a minimal-uncertainty wavepacket. For the general case, one can identify $l_z = \hbar/\sigma_p$ and $l_p = \hbar/\sigma_z$ as the relevant scales for coherence [26, 29], where l_z and l_p are the spatial coherence length and the momentum coherence width, respectively.

Let us discuss the meaning of this equation. The quantities σ_z and σ_p characterize the initial atomic wavefunction, and are thus microscopic quantities. The quantities Δz and Δp_z describe the experimental imprecision in the final recombination. In a "good" SG experiment (i.e., one which allows "unmistakable" splitting [29]) the maximum values of splitting in position and momentum should be much larger than their respective initial widths, meaning they should be macroscopic. On the other hand, according to Eq. (2), a nearly perfect maintenance of spin coherence ($V \simeq 1$) requires both $\Delta z \ll \sigma_z$ and $\Delta p_z \ll \sigma_p$. Consequently, Eq. (2) tells us that we need to recombine macroscopic quantities with a microscopic level of precision. This is the challenge facing SG interferometer experiments.

It is interesting to note that in the half-loop experiments, we found that Δp_z can be quite large (rendering the trajectories during the TOF period in Fig. 2 slightly non-parallel) without significantly reducing the measured spatial interference fringe visibility, so the stability of the half-loop experiments cannot be used to examine the HD equation. This robustness of the half-loop may also be understood by considering the fact that the expansion of the wavepackets creates an enhanced local coherence length, since for every region of space the k vector variance becomes smaller as TOF increases (see also [69, 70]).

A practical full-loop SG experiment must consider and address two effects. First, as noted above, the HD effect requires accurate recombination, namely, small Δz and Δp_z. These small values must be maintained for many experimental cycles, and thus a high level of stability in these values is also important. Achieving accurate recombination means that the overlap integral, calculated in Eq. (2), will have a significant non-zero value. Second, one must maintain a stable interferometer phase $\delta\Phi$, so that it has the same value shot-to-shot. This requires that the coupling to external magnetic noise is kept to a minimum, either by shielding the experiment and stabilizing the electronics (e.g., responsible for the homogeneous magnetic fields), or by conducting the experiment extremely quickly so that such environmental fluctuations do not have time to introduce significant phase noise.

Our full-loop SGI yields a visibility up to 95% (Fig. 4d), proving that we are able to use the SG effect to build a full-loop interferometer as originally envisioned almost a century ago. We note three differences between our realization and the scheme considered in the HD papers: (1) We use a BEC, which is a minimum-uncertainty wavepacket, whereas the HD papers considered atomic beam experiments with large uncertainties on the order of $\sigma_z\sigma_p \simeq 10^3$; (2) We implement fast magnetic gradient pulses generated by running currents on the atom chip, in contrast to using constant gradients from permanent magnets that were considered in the original proposals; (3) Our interferometer is a 1D longitudinal interferometer, while the originally envisioned SGI was 2D, i.e., it enclosed an area.

The full-loop experiments include a wide variety of optimizations and checks (see [64] for additional details). To make sure the spin superposition is not dephased due to some slowly varying gradients in our bias fields, we add π pulses giving rise to an echo sequence. To access a larger region of parameter space and to ensure the robustness of our results, we use several different configurations by, for example, implementing the reversing pulse (T_3) by inverting the sign of the atom chip currents vs. inverting the spins with the help of π pulses. We also utilize a variety of magnetic gradient magnitudes, and scan both the splitting gradient pulse duration T_1 and the delay time between the pulses T_d. All results are qualitatively the same. For weak splitting we observe high visibility ($\sim$95%), while for a momentum splitting equivalent to $\hbar k$ the visibility is still high ($\sim$75%), indicating that the magnet precision enabled coherent spin-state recombination to a high degree.

Finally, we briefly compare our experiments to previous work in an elaborate series of SGI experiments over a period of 15 years using metastable atomic beams [35–42, 44] and, more recently, thermal and ultra-cold alkali atoms [43, 45]. A detailed discussion is given in [64]. While these longitudinal beam experiments did observe spin-population interference fringes, the experiments reviewed here are very different. Most importantly, an analogue of the full-loop configuration was never realized, as only splitting and stopping operations were applied (i.e., there was no recombination) and wavepackets emerged from the interferometer with the same separation as the maximal separation achieved within (see Fig. 2 of [40] and Footnote [10] of [43]). We have not found anywhere in the many papers published by this group (only some of which are referenced here) evidence of four operations being applied as required for a full-loop configuration, whether the experiment was with longitudinal or transverse gradients. In addition, no spatial interference fringes were observed, as the spatial modulation they observed was a signature of multiple parallel longitudinal interferometers, each having its own individual relative phase between its two wavepackets.

To conclude, we have shown that a full-loop has been realized [18, 121]. In addition, as previously shown in Heisenberg's argument, the momentum splitting is the figure of merit in determining the phase dispersion. In our experiment, coherence is observed up to a momentum splitting as high as $\Delta p_z(T_1)/\sigma_p = 60$. However, in contrast, the visibility is more sensitive to spatial splitting and we achieve $\Delta z(T)/\sigma_z = 4$, much lower than for the half-loop, where we achieved $\Delta z/\sigma_z = 18$. The splitting is coherent but its, limits in terms of the HD effect are yet to be explored quantitatively. Many mysteries remain to be solved, such as why is the observed reduction not symmetric in momentum and spatial splitting, in contrast to Eq. 2. A simple answer, which is yet to be examined in detail, is the existence of some sort of spatial decoherence mechanism due to the environment.

Having now described the SG beam-splitter, the SG half-loop, and the SG full-loop, we show in the next section how these techniques may be used for different applications.

6   Applications

The pulse sequence in the half-loop experiments creates two spatially separated wavepackets in the state $|2\rangle$ with zero relative momentum (left-most frame of Fig. 5a–c). We now take advantage of the long free-fall period in the experiment (labelled TOF in Fig. 2, i.e., after the "stopping pulse") to further manipulate these wavepackets while they are allowed to expand and ultimately to overlap. The experiments are based on imposing a differential time evolution between the two wavepackets, which we measure as the interference patterns generated upon their recombination.

In particular, we create a "clock" state for each of the two wavepackets by first applying an RF pulse that prepares the atoms in a superposition of two Zeeman sublevels $|1\rangle$ and $|2\rangle$ whose coefficients depend on the Bloch sphere angles θ and ϕ. This superposition state is a two-level system evolving with a known period, as in the regular notion of an atomic clock. The RF pulse (duration T_R) controls the value of $C = \sin\theta$, while a subsequent magnetic gradient pulse (duration T_G) controls the value of $D_I = \sin(\phi/2)$ by changing the relative "tick" rate $\Delta\omega$ of the two clock wavepackets, as illustrated in Fig. 5a–c. The quantities C and D_I describe the clock preparation quality and the ideal distinguishability between the two clock interferometer arms respectively, which we will find quantitatively useful in our discussion of clock complementarity [see Eqs. (4) and (5)]. We note that, although the magnetic gradient pulse applies a different SG force to each of the states within the clock, we have evaluated this effect for our experimental parameters and find that it is smaller than our experimental error bars ($\leq 2\%$, Supplementary Materials of [21]).

6.1 Clock Interferometery

Let us first discuss the motivation for clock interferometry [20]. Time in standard quantum mechanics (QM) is a global parameter, which cannot differ between paths. Hence, in standard interferometry [71], a height difference in a gravitational field between two paths would merely affect the relative phase of the clocks, shifting the interference pattern without degrading its visibility. In contrast, general relativity (GR) predicts that a clock must "tick" slower along the lower path; thus if the paths of a clock passing through an interferometer have different heights, a time differential between the paths will yield "which path" information and degrade the visibility of the interference pattern according to the quantum complementarity relation between the interferometric visibility and the distinguishability of the wavepackets [72]. Consequently, whereas standard interferometry may probe GR [73–75], clock interferometry probes the interplay of GR and QM. For example, loss of visibility because of a proper time lag would be evidence that gravitational effects contribute to decoherence and the emergence of a classical world [76].

Here we describe the use of this new tool—the clock interferometer—for its potential to investigate the role of time at the interface of QM and GR. Since the genuine GR proper time difference is too small to be measured with existing experimental technology, our experiments instead simulate the proper time difference between the clock wavepackets using magnetic gradients, thereby causing the clock wavepackets to "tick" at different rates. Our results in this proof-of-principle experiment show that the visibility does indeed oscillate as a function of the simulated proper time lag.

In the ultimate experiment, each part of the spatial superposition of a clock, located at different heights above Earth, would "tick" at different rates due to gravitational time dilation (so-called "red-shift"). We can easily calculate the proper time difference between two arms of the clock interferometer as a figure-of-merit for this effect. Using a first-order approximation of gravitational time dilation, and assuming a large separation between the arms of $\Delta h = 1$ m, an interferometer duration of $T = 1$ s yields a proper time difference between the arms of only $\Delta \tau \simeq T g \Delta h / c^2 \simeq 10^{-16}$ s. Such a small time difference means that a very accurate and fast-ticking clock must be sent through an interferometer with a large space-time area in order to observe the actual GR effect. Both requirements are beyond our current experimental capabilities. Our "synthetic" red-shift is created by applying an additional magnetic gradient

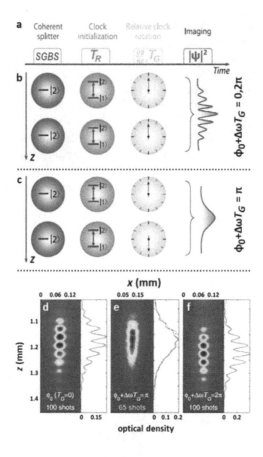

Fig. 5 Clock interferometry. **a** Timing sequence (not to scale): Following a coherent spatial splitting by the SGBS and a stopping pulse, the system consists of two wavepackets in the $|2\rangle$ state (separated along the z axis) with zero relative velocity, as in Sect. 4. The clock is then initialized with an RF pulse of duration T_R (usually a $\pi/2$ pulse, $T_R = 10\,\mu s$) after which the relative "tick" rate $\Delta\omega$ of the two clock wavepackets may be changed by applying a magnetic field gradient $\partial B/\partial z$ of duration T_G. Clock initialization occurs 1.5 ms after trap release, the first 0.9 ms of which is used for preparing the two wavepackets. The wavepackets are then allowed to expand and overlap and an image is taken. **b** Evolution in time, synchronized with (**a**). Each ball represents a clock wavepacket, where the hand represents its Bloch sphere phase ϕ_{BS}. When the clock reading (i.e., the position of the clock hand) in the two clock wavepackets is the same ($\phi_{BS} = \phi_0 + \Delta\omega T_G = 0, 2\pi$), fringe visibility is high. **c** When the clock reading is opposite (orthogonal, $\phi_{BS} = \phi_0 + \Delta\omega T_G = \pi$), there is no interference pattern. (d)-(f) Corresponding interference data of the two wavepackets, i.e., of the clock interfering with itself. All data samples are from consecutive measurements without any post-selection or post-correction. Single-shot patterns for $\phi_{BS} = \phi_0 + \Delta\omega T_G = \pi$ also show very low fringe visibility (see Fig. 2c of [20]). Adapted from [20] and reprinted with permission from AAAS; **e** is adapted from [64]

(of duration T_G) that causes the clock wavepackets to "tick" at different rates. We denote the "tick" rate difference by $\Delta\omega$.

Our results, some of which are presented in Fig. 5d–f, with more details in [20, 64], show that the relative rotation between the two clock wavepackets affects the interferometric visibility. In the most extreme case, when the two clock states are orthogonal, e.g., one in the state $\frac{1}{\sqrt{2}}(|1\rangle + |2\rangle)$ and the other in the state $\frac{1}{\sqrt{2}}(|1\rangle - |2\rangle)$, the visibility of the clock self-interference drops to near zero (Fig. 5e). By varying the duration of the magnetic gradient T_G and thereby scanning the differential rotation angle ϕ_{BS} between the two clock wavepackets, we show quantitatively that the visibility oscillates as a function of our "synthetic" red-shift with a period of $\Delta\omega T_G = 2\pi$ (Fig. 5d,f). As an additional test of the clock interferometer, we modulate its preparation by changing the duration of the clock initialization pulse T_R, which influences the relative populations of the two states composing the clock. This changes the state of the system from a no-clock state to a full-clock state in a continuous manner. The results show that the visibility behaves as expected in each case, further validating that it is the clock reading which is responsible for the oscillations in visibility that we observe as a function of T_R [20].

6.2 Clock Complementarity

These measurements of visibility may naturally be extended to study quantum complementarity for our self-interfering atomic clocks, which we again remark is at the interface of QM and GR. Our central consideration here is the inequality [77]

$$V^2 + D^2 \leq 1, \tag{3}$$

where V is the "visibility" of an interference pattern such as discussed throughout this review, and D is the "distinguishability" of the two paths of the interfering particle. The latter quantity can also be measured directly in the clock experiments by controlling the angle ϕ_{BS}, where $(\theta = \pi/2, \phi_{BS} = \Delta\omega T_G = \pi)$ prepares two perfectly distinguishable clocks such that $D = 1$ (Fig. 5e). A brief account of recent work theoretically and experimentally verifying this fundamental inequality is given by [21] and references therein.

It is important to investigate clock complementarity, particularly in view of recent theoretical work showing that spatial interferometers can be sensitive to a proper time lag between the paths [78] and speculation (see Table 1 in [72]) that the inequality of Eq. (3) may be broken such that $V^2 + D^2 > 1$ when the effect of gravity is dominant. Zhou et al. summarize the importance of this work as follows: "... on the one hand, if the 'ticking' rate of the clock depends on its path, then clock time provides which-path information and Eq. (3), developed in the framework of non-relativistic QM, must apply. Yet, on the other hand, gravitational time lags do not arise in non-relativistic QM, which is not covariant and therefore not consistent with the equivalence principle [79]. Hence our treatment of the clock superposition is a semiclassical extension of quantum mechanics to include gravitational red-shifts."

The experiments we conducted in [21] set out to test Eq. (3) quantitatively. Imperfect clock preparation (i.e., with $\theta \neq \pi/2$) reduces the measurable distinguishability D from its ideal value D_I as

$$D^2 = (C \cdot D_I)^2, \quad \text{where } C \equiv \sin \theta = 2\sqrt{P(1-P)} \qquad (4)$$

and

$$D_I = |\sin(\Delta\phi/2)| \qquad (5)$$

with P and $1 - P$ denoting the populations (occupation probabilities) of the two energy eigenstates of the clock and $\Delta\phi_{BS} \equiv \phi_{BS}^u - \phi_{BS}^d$, where u and d denote the upper and lower paths of the interferometer, respectively.

The experiment now has the task of measuring the three quantities V, D_I and C independently. We use the normalized visibility V_N as discussed in Sect. 4. We evaluate D_I independently by measuring the relative phases in two single-state interferometers, one for each of the two clock states, and we measure C, also independently, in a separate experiment by measuring P after the clock is initialized. Our results for these independently-measured quantities are shown in Fig. 6a, c, and e, where the results in (c) and (e) are based on analyzing the data in (b) and (d) respectively. We then combine these three quantities in the complementarity expression

$$(V_N)^2 + (C \cdot D_I)^2 \leq 1, \qquad (6)$$

whereupon we see from Fig. 6f–g that the complementarity inequality [Eq. (3)] is indeed upheld for the clock wavepackets superposed on two paths through our SG interferometer.

While the relation in Eq. (6) is specific to clock complementarity, it is unusual in linking non-relativistic quantum mechanics with general relativity. A direct test of this complementarity relation will come when D_I reflects the gravitational red-shift between two paths which traverse different heights.

6.3 Geometric Phase

The geometric phase due to the evolution of the Hamiltonian is a central concept in quantum physics. In noncyclic evolutions, a proposition relates the geometric phase to the area bounded by the phase-space trajectory and the shortest geodesic connecting its end points [80–82]. The experimental demonstration of this geodesic rule proposition in different systems is of great interest, especially due to its potential use in quantum technology. Here, we report a novel experimental confirmation

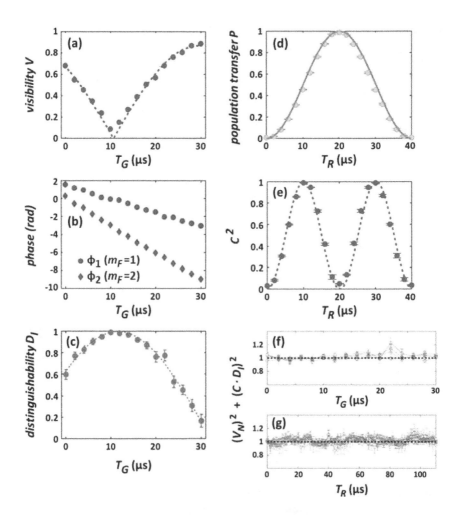

Fig. 6 Clock complementarity: **a–e** V, D_I, and C measured independently and **f–g** combined in the complementarity relations of Eqs. (3)-(6). **a** The visibility of an ideal clock ($C = 1$) interference pattern vs. T_G, fitted to $|\cos(\phi/2)|$; **b–c** the distinguishability is calculated from Eq. (5) using the difference in relative angles $\phi_2 - \phi_1$, each measured separately and shown in (**b**); and **d–e** the clock preparation quality C is calculated from Eq. (4) using the data in (**d**). Finally, **f** shows the combination of all three parameters $(V_N)^2 + (C \cdot D_I)^2$ for four values of C when D_I is scanned and **g** shows the same combination for four values of D_I when C is scanned. Only the data point in (**f**) for T_G near 22 μs differs from unity, due to a relatively large experimental error in measuring the interferometric phase. These data therefore verify clock complementarity. Adapted from [21] with permission © IOP Publishing & Deutsche Physikalische Gesellschaft, all rights reserved

of the geodesic rule for a noncyclic geometric phase by means of a spatial SU(2) matter-wave interferometer, demonstrating, with high precision, the predicted phase sign change and π jumps. We show the connection between our results and the Pancharatnam phase [83].

In the clock complementarity application just described, we scanned the third RF pulse (duration T_R) to vary the clock preparation parameter $C = \sin\theta$. In our case, a $\pi/2$ pulse typically corresponds to $T_R = 10\,\mu s$, so $T_R < 10\,\mu s$ places the Bloch vector in the northern hemisphere of the Bloch sphere with $P_1 < P_2$, while $10 < T_R < 30\,\mu s$ places the Bloch vector in the southern hemisphere ($P_1 > P_2$), i.e., the selected hemisphere is a periodic function of T_R such that an unequal superposition of $|1\rangle$ and $|2\rangle$ is created for each of the wavepackets unless θ lies on the equator. After applying this RF pulse (with some chosen duration T_R), we adjust the phase difference between the two superpositions by applying the third magnetic gradient pulse of duration T_G. This rotates the Bloch vectors along the latitude that was selected by the RF pulse to points A and B in the northern hemisphere (or A', B' in the southern hemisphere) as shown in Fig. 7a, thereby affecting the phase difference $\Delta\phi_{BS}$, which we simply call $\Delta\phi$ hereafter.

The two wavepackets are allowed to interfere as in our half-loop experiments, enabling a direct measurement of the geometric phase. As usual, we extract the "total" interference phase (labeled Φ) by fitting the fringe patterns using Eq. (1). For general values of θ and $\Delta\phi$ (i.e., after the application of both T_R and T_G), we write the total phase between the two wavepackets as [84]

$$\Phi = \arctan\left\{\frac{\sin^2(\theta/2)\,\sin(\Delta\phi)}{\cos^2(\theta/2) + \sin^2(\theta/2)\,\cos(\Delta\phi)}\right\}. \tag{7}$$

Measurements of Φ, combined with values of θ deduced independently from the relative populations of states $|1\rangle$ and $|2\rangle$, then allow us to fit $\Delta\phi$ to high precision as a function of T_G. These measurements verified that $\Delta\phi$ depends linearly on T_G, and we found that $\Delta\phi = \pi$ occurs at $T_G = 17\,\mu s$.

Figure 7b–c shows interference fringe images for this specific value of T_G, from which we extract the total phase as shown in Fig. 7e. We see immediately that this phase is independent of θ within each hemisphere, an observation we call "phase rigidity". Moreover, the (constant) phase in each hemisphere differs by π, which can also be deduced from the vanishing visibility shown in Fig. 7d in which we have combined the data from both hemispheres. Evidently, there is a sharp jump in the phase of the interference pattern as θ crosses the equator, as suggested by the singularities in Eq. (7) that arise when $\theta = \pi(n + 1/2)$ (integer n) and $\Delta\phi = \pi$.

To understand the non-cyclic geometric phase, we need to further examine the Bloch sphere. We see that the path from $A \to B$ along the latitude θ and returning along the geodesic (or "great-circle route") from $B \to A$ encloses an area [blue shading in Fig. 7(a)] in a counter-clockwise direction, whereas the corresponding path from $A' \to B'$ and back again in the southern hemisphere proceeds in a clockwise direction. One-half of this area is the "geometric phase" that we now wish to calculate.

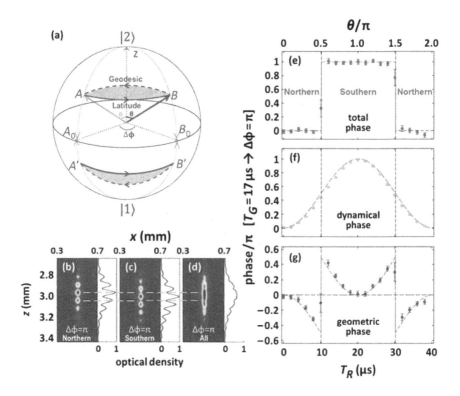

Fig. 7 Geometric phase. **a** Bloch sphere for the two wavepackets (green and red arrows labeled A and B, respectively) prepared by an RF pulse (duration T_R, rotation angle θ) and a subsequent magnetic gradient pulse (duration T_G) that induces a rotation angle difference of $\Delta\phi$. The rotation $A \rightarrow B$ lies along a constant latitude (solid purple line), while the returning geodesic $B \rightarrow A$ lies along the "great circle" curve (dashed purple line). Bloch vectors for corresponding wavepackets prepared in the southern hemisphere are shown as A' and B'. **b–c** Interference fringes generated by the half-loop SGI, averaged over a total of 330 experimental shots with varying $0 < T_R < 40 \,\mu s$, while keeping a fixed value of $T_G = 17 \,\mu s$ (this value of T_G corresponds to $\Delta\phi = \pi$, see text). The dashed green lines show that the maxima in (**b**) lie exactly where the minima occur in (**c**), corresponding to Bloch vectors prepared in the northern and southern hemispheres, respectively. Adding all these interference patterns together in (**d**) shows near-zero visibility, i.e., they are completely out of phase. The fact that exactly the same pattern is observed while in the same hemisphere, independent of θ (duration of T_R), is called "phase rigidity". **e** Total phase extracted from the interference fringes measured as a function of the RF pulse duration (lower scale) and the corresponding latitude θ (upper scale). Phase rigidity is clearly visible. **f–g** Dynamical and geometric phases extracted from the data in (**e**) and independently measured values of θ and $\Delta\phi$ (see text). The range of T_R in (**e–g**) (T_G is fixed at $17 \,\mu s$) corresponds to a full cycle from the northern hemisphere ($0 < T_R < 10 \,\mu s$) through the southern hemisphere ($10 < T_R < 30 \,\mu s$), and back to the north pole at $T_R = 40 \,\mu s$. Adapted from [84] with permission © the authors, some rights reserved; exclusive licensee AAAS. CC BY 4.0

The total phase change Φ for closed paths like $A \rightarrow B \rightarrow A$ and $A' \rightarrow B' \rightarrow A'$ is a sum of two contributions, the dynamical phase Φ_D and the geometric phase Φ_G. The dynamical phase is given by [80]

$$\Phi_D = \frac{\Delta\phi}{2} \left(1 - \cos\theta\right), \tag{8}$$

which can be determined by measuring θ and $\Delta\phi$ independently. For the particular value of $\Delta\phi = \pi$ chosen as a sub-set of our experimental data, we are then able to present Φ_D in Fig. 7f. Finally, we subtract the phases Φ_D, as plotted in (f), from the total phases Φ plotted in (e) (which, as noted above, are extracted directly from the observed interference pattern) to obtain the phases Φ_G. Namely, we perform $\Phi - \Phi_D$ and get Φ_G, which is presented in Fig. 7g. Let us emphasize that the total phase Φ is also the Pancharatnam phase [83], and thus our experiment is also a direct measurement of this phase.

Our plot of Φ_G exactly confirms the prediction shown in Fig. 4d of [81], also reproduced as the dashed blue line in Fig. 7g. The predicted sign change as the latitude crosses the equator is clearly visible. The evident phase jump is due to the geodesic rule. When $\Delta\phi = \pi$, the geodesic must go through the Bloch sphere pole for any $\theta \neq \pi/2$. As the latitude approaches the equator (i.e., increasing θ), the blue area in Fig. 7a (twice Φ_G) grows continuously, reaching a maximum of π in the limit as $\theta \rightarrow \pi/2$. As the latitude crosses the equator, the geodesic jumps from one pole to the other pole, resulting in an instantaneous change of sign of this large area and a phase jump of π.

Finally, our approach for testing the geodesic rule is unique for the following reasons: (1) the use of a spatial interference pattern to determine the phase in a single experimental run (no need to scan any parameter to obtain the phase); (2) the use of a common phase reference for both hemispheres while scanning θ, enabling verification of the π phase jump and the sign change; and (3) obtaining the relative phase by allowing the two coherently-prepared wavepackets to expand in free flight and overlap, in contrast to previous atom interferometry studies that required additional manipulation of θ and $\Delta\phi$ to obtain interference.

6.4 T^3 Stern-Gerlach Interferometer

Here we consider an application of the full-loop SGI wherein we minimize the delay times between successive SG pulses as much as allowed by our electronics. In such an extreme scenario, it is expected that the phase accumulation will scale purely as T^3, thus representing the first pure interferometric measurement of the Kennard phase [19] predicted in 1927 [85, 86] (see also [87–89]). The theory for this experiment was done by the group of Wolfgang Schleich.

In order to describe the phase evolution of an atom moving in a time- and state-dependent linear potential, it is sufficient [90] to know the two time-dependent forces $\mathbf{F}_u \equiv F_u(t)\mathbf{e}_z$ and $\mathbf{F}_l \equiv F_l(t)\mathbf{e}_z$ acting on the atom along the upper and lower branches, respectively, of the interferometer shown in Fig. 8, where z is the axis of gravity, the axis of our longitudinal interferometer, and also the axis of our magnetic gradients.

In the present case, these forces comprise the gravitational force $F_g = mg$ and the state-dependent magnetic forces $\mathbf{F}_i = -\mu_B(g_F)_i(m_F)_i(\partial|\mathbf{B}|/\partial z)\mathbf{e}_z$, $(i = 1, 2)$:

$$F_{u,l}(t) = F_g + F_{2,1}\mathcal{F}(t), \tag{9}$$

where μ_B, g_F, and m_F are the Bohr magneton, the Landé factor, and the projection of the angular momentum on the quantization (y-)axis, respectively. The function $\mathcal{F}(t)$ provides the time-dependent modulation shown as the orange curve in Fig. 8(b):

$$\mathcal{F}(t) \equiv \Theta(t) - \Theta(t - T_1) - \Theta(t - T_1 - T_d) + \Theta(t - 3T_1 - T_d)$$
$$+ \Theta(t - 3T_1 - 2T_d) - \Theta(t - 4T_1 - 2T_d). \tag{10}$$

Here we are using the Heaviside step function $\Theta(t)$ and we are assuming that the duration of each gradient pulse is identical, i.e., $T_{2,3,4} = T_1$, as are the two delay times, $T_{d_1,d_2} = T_d$. We are also careful to ensure experimentally that the magnetic field is linear in the vicinity of the atoms and acts only along the vertical (z-)axis.[3]

As in the full-loop SGI experiments of Sect. 5, we measure the spin population in state $|1\rangle$ which, in this configuration, is a periodic function of the interferometer phase [91].

$$P_1 = \frac{1}{2}[1 - \cos(\delta\Phi + \varphi_0)], \tag{11}$$

where

$$\delta\Phi = \frac{1}{\hbar}\int_0^T dt\, \bar{F}(t)\delta z(t), \tag{12}$$

with the total time $T \equiv 4T_1 + 2T_d$. Note that the interferometer will be closed in both position and momentum provided that the differences

$$\delta p(t) = \int_0^t d\tau\, \delta F(\tau) \tag{13}$$

---

[3]Magnetic field linearity is ensured to a good approximation by the three-wire chip design and by carefully positioning the atoms very close to the center of the quadrupole field that they produce, as well as by the short distances that the atomic wavepackets travel ($\sim$1 μm) compared to their distance from the chip ($\sim$100 μm). We also adjust the duration of T_4 slightly, relative to T_1, to better optimize the visibility and account for any residual non-linearity. See [19, 68] for further details.

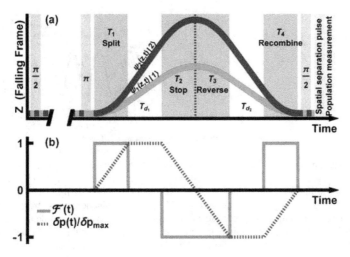

Fig. 8 Pulse sequence of our longitudinal T^3-SGI (not to scale). **a** Trajectories of the atomic wavepackets with internal states $|1\rangle$ (green curve) and $|2\rangle$ (purple curve). Here we are using the freely-falling reference frame (gravity upwards), distinct from the center-of-mass reference frame used for Figs. 2 and 4. Also shown are the RF (blue) and magnetic gradient (red) pulses. The magnetic field gradients result in a state-dependent force along the z-direction while the strong bias magnetic field along the y-direction defines the quantization axis and ensures a two-level system. **b** Time dependence of the relative force $\mathcal{F} = \mathcal{F}(t)$ [orange curve, Eq. (10)] and the corresponding relative momentum $\delta p(t)$ [blue dashed curve, Eq. (13)] between the wavepackets moving along the two interferometer paths. In the experiment, we achieved the maximal separation $\Delta z_{\max} = 1.2\,\mu\text{m}$ in position and $\Delta p_{\max}/m_{\text{Rb}} = 17\,\text{mm/s}$ in velocity. Reprinted from [19] with permission © (2019) by the American Physical Society

and

$$\delta z(t) = \frac{1}{m} \int_0^t d\tau\, \delta F(\tau)(t - \tau) \tag{14}$$

both vanish at $t = T$. Here φ_0 is a constant phase taking into account possible technical misalignment, while $\bar{F}(t) \equiv [F_u(t) + F_l(t)]/2 = F_g + \frac{1}{2}(F_1 + F_2)\mathcal{F}(t)$ and $\delta F(t) \equiv F_u(t) - F_l(t) = (F_2 - F_1)\mathcal{F}(t)$ are the mean and relative forces respectively. From Eq. (11) we finally obtain

$$\begin{aligned}
\delta\Phi &= \frac{mga_B}{\hbar}\left(\frac{\mu_1 - \mu_2}{\mu_B}\right)\left(2T_1^3 + 3T_1^2 T_d + T_1 T_d^2\right) \\
&\quad + \frac{ma_B^2}{\hbar}\left(\frac{\mu_1^2 - \mu_2^2}{\mu_B^2}\right)\left(\frac{2}{3}T_1^3 + T_1^2 T_d\right),
\end{aligned} \tag{15}$$

with $a_B \equiv \mu_B \nabla B / m$ being the magnetic acceleration.

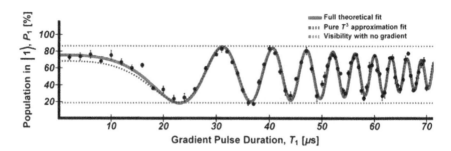

Fig. 9 Measurement of the cubic phase with the T^3-SGI presented in Fig. 8. The solid red line represents a fit based on Eq. (15), as described in the text. The dashed blue line is a fit with $T_d = 0$, showing that the interferometer phase scales purely as T_1^3 for $T_1 \gtrsim 20\,\mu s$. The visibility drops from 68% to 32% over 70 μs with a decay time of 75 μs. This reduction results from inaccuracies in recombining the two interferometer paths. The dashed gray horizontal lines depict the maximal and minimal values of the population P_1 measured independently without magnetic field gradients. Reprinted from [19] with permission © (2019) by the American Physical Society

As sketched in Fig. 8, the experiment begins with an on-resonance RF $\pi/2$-pulse that transfers the initially prepared internal atomic state $|2\rangle$ to an equal superposition, $\frac{1}{\sqrt{2}}(|1\rangle + |2\rangle)$. This $\pi/2$ pulse is applied 1 ms after the atoms are released from the trap in which they were prepared, in order to ensure that the trapping fields are fully quenched. Following a free-fall time of 400 μs (the first "dark time"), we apply an RF π-pulse that flips the atomic state to $\frac{1}{\sqrt{2}}(|1\rangle - |2\rangle)$. After a second dark time of another 400 μs, a second $\pi/2$ pulse completes the spin-echo sequence. The π-pulse inverts the population between the two states of the system thereby allowing any time-independent phase shift accumulated during the first dark time to be canceled in the second dark time. The experiment is completed by applying a magnetic gradient to separate the spin populations and a subsequent pulse of the detection laser to image both states simultaneously.

As with all our previous full-loop experiments, the four magnetic field gradient pulses are produced by current-carrying wires on the atom chip. This magnetic pulse sequence sends the spin states $|1\rangle$ and $|2\rangle$ along different trajectories in the SGI and ultimately closes the interferometer in both momentum and position. Careful calibration measurements verified that reversing the wire currents (the current flow is reversed during T_2 and T_3 relative to T_1 and T_4) provides magnetic accelerations that are equal in magnitude (but opposite in sign) to within our experimental uncertainty of $< 1\%$.

The experimental data shown in Fig. 9 are measured as a function of the time $2 < T_1 < 70\,\mu s$. From Eq. (15), it is apparent that the T^3 dependence will be most evident if $T_d \ll T_1$, which is satisfied for most of the experimental range by using a fixed experimental value of $T_d = 2.6\,\mu s$ (limited by the speed of our electronic circuits). Note that $T_1 \lesssim 100\,\mu s$ is limited by the duration of the second dark time.

The experimental data (dots) agree very well with the theory (solid red line) based on Eq. (15), where the fitting parameters are the magnetic acceleration a_B as well as the decay constant of the visibility and a constant phase φ_0. The dashed blue line is obtained by setting $T_d = 0$, leading to a pure T_1^3 scaling that is indistinguishable from the full theoretical fit for $T_1 \gtrsim 20\,\mu s$:

$$\delta\Phi^{(T^3)} \cong \frac{ma_B}{32\hbar}\left(\frac{\mu_1 - \mu_2}{\mu_B}\right)\left(g + \frac{\mu_1 + \mu_2}{3\mu_B}a_B\right)T^3. \tag{16}$$

The maximum visibility displayed by the gray lines is first measured by performing only the RF spin-echo sequence $(\pi/2 - \pi - \pi/2)$ without the magnetic field gradients and changing the phase of the second $\pi/2$ pulse. The maximal visibility is limited by imperfections in the RF pulses. As discussed above, utilizing an echo sequence allows us to cancel out contributions to the interferometer phase from the bias magnetic field, and to increase the coherence time.

The excellent fit to these data allows a precise determination of the magnetic field acceleration, $a_B^{\mathrm{fit}} = 246.97 \pm 0.09\,\mathrm{m/s^2}$. Separate measurements were used to independently determine the magnetic field gradient using time-of-flight (TOF) techniques, which gave a value of $a_B^{\mathrm{TOF}} = 249 \pm 2\,\mathrm{m/s^2}$.[4] While these measurements agree with one another, the difference in measurement errors clearly shows that our T^3-SGI provides a much more precise measurement of the magnetic field gradient.

Let us now consider the case when $T_1 \ll T_d$, such that during T_d the relative momentum $\delta p_0 \equiv ma_B T_1(\mu_1 - \mu_2)/\mu_B$ between the paths is kept constant, i.e., we take the magnetic field gradient pulses to be delta functions.

In this limit the interferometer phase from Eq. (15) becomes

$$\delta\Phi^{(T^2)} \cong \frac{\delta p_0}{4\hbar}gT^2, \tag{17}$$

scaling quadratically with the total time $T \cong 2T_d$, since we now maintain a piecewise constant momentum difference between the two arms. This is similar to the T^2-SGI [18] or the Kasevich-Chu interferometer [90], although the momentum transfer δp_0 is provided by the magnetic field gradient in the case of the T^2-SGI, rather than by the laser light pulse.

We conclude our discussion of this unique T^3 interferometer by comparing the scaling of the interferometer phases $\delta\Phi^{(T^3)}$ and $\delta\Phi^{(T^2)}$ with the total interferometer time T, as given by Eqs. (16) and (17) respectively. The data in Fig. 10 are taken from Fig. 9 and from our T^2-SGI (when experimentally realizing the condition $T_1 \ll T_d$), showing clearly that the T^3-SGI significantly outperforms the T^2-SGI with respect to total phase accumulation, even though the latter can currently operate for total times T up to three times longer than the former. Finally, let us briefly note that this T^3 realization has already been coined a proof-of-principle experiment for testing the quantum nature of gravity [49].

---

[4]These values for a_B^{fit} and a_B^{TOF} are different from those presented in [19] due to a different fitting procedure used there. A full analysis and fitting procedures are presented in the Appendices of [68].

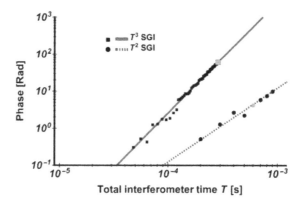

Fig. 10 Scaling of the interferometer phases $\delta\Phi^{(T^3)}$ [squares, Eq.(16)] and $\delta\Phi^{(T^2)}$ [circles, Eq. (17)], as functions of the total interferometer time T. The solid red line is fitted to our data for the T^3-SGI and the dashed blue line is fitted to our T^2-SGI data when experimentally realizing the condition $T_1 \ll T_d$. In its current configuration with $T_{max} = 285\,\mu s$, the phase of the T^3-SGI is almost six times larger than the phase of the best T^2-SGI, even though the magnetic field gradients and the maximal time $T_{max} = 924\,\mu s$ are larger than those of the T^3-SGI by factors of 2.3 and 3.2, respectively. For reference, the green square and green dot represent data for which the observed visibility is $\approx 30\%$ for both the T^3-SGI and T^2-SGI respectively. Adapted from [19] with permission © (2019) by the American Physical Society

Looking into the future, we may ask if one may extend the T^3 scaling to yet higher powers of time. In the Ramsey-Bordé interferometer [92], the phase shift that scales linearly with the interferometer time T originates from a constant position difference between two paths during most of this time. In the Kasevich-Chu interferometer [93, 94], the quadratic scaling of the phase with time is caused by a piecewise constant velocity difference, while a piecewise constant acceleration difference between the two paths results in the cubic phase scaling $\delta\Phi \propto T^3$, as presented above.

One can generalize this idea to achieve any arbitrary phase scaling by having a piecewise difference in the nth derivative of the position difference between the two paths. By designing an interferometer sequence consisting of pulses with a higher-order time-dependence of the forces, combined with careful choices of the relative signs and durations of the pulses, the total phase can be made to scale with the interferometer time as T^{n+1} for any chosen $n > 2$.

7 Outlook

7.1 SGI with Single Ions

The discovery of the Stern-Gerlach effect led to lively discussions early in the quantum era regarding the possibility of measuring an analogous effect for the electron itself (see e.g., [95, 96]). The Lorentz force adds the complicating factor of a purely classical deflection of the electron beam that would smear out any expected SG splitting. Here we summarize a generalized semiclassical discussion for any charged particle of mass m and charge e from [62] (though with the co-ordinate system in Table 14.1). Assuming a beam momentum p_x and a transverse beam spatial width Δz, we calculate the spread of the Lorentz force ΔF_L due to a transverse magnetic gradient B' as

$$\Delta F_L = \frac{e}{m} \, p_x B' \Delta z. \tag{18}$$

Since the beam would be well collimated, $\Delta p_z < p_x$, so

$$\Delta F_L > \frac{e}{m} \, B' \Delta p_z \Delta z \geq \frac{e\hbar}{2m} \, B' = \frac{m_e}{m} \times \left(\frac{e\hbar}{2m_e} \, B' \right) = \frac{m_e}{m} \, F_{SG}, \tag{19}$$

where the second inequality uses the uncertainty principle and we have introduced the electron mass m_e to relate F_L to the Stern-Gerlach force F_{SG}.

The spatially inhomogeneous Lorentz broadening is therefore larger than the SG splitting for electrons ($m = m_e$), at least in this semiclassical analysis [97], and this lively controversy has continued for decades though, as far as we know, without any conclusive experimental tests for electrons or for any other charged particles (see [98–100] for reviews of the early history of this issue and recent perspectives). In contrast, Eq. (19) shows no such fundamental problem if we take ions such that $m_e/m < 10^{-3}$, thereby motivating our proposals, including chip-based designs, for measurements using very high-resolution single ion-on-demand sources that have recently been developed using ultra-cold ion traps [61, 101]. As a practical matter, we note that a suitable ion chip could be fabricated and implemented either based on an array of current-carrying wires as analyzed in [62] or on a magnetized microstructure like those implemented in [102, 103].

Although we did not extend our analysis to include the coherence of the spin-dependent splitting, the suggested ion-SG beam splitter may form a basic building block of free-space interferometric devices for charged particles. Here we quote from our collaborative work with Henkel et al. [62]. In addition to measuring the coherence of spin splitting as in the "Humpty-Dumpty" effect (see Sect. 1), we anticipate that such a device could provide new insights concerning the fundamental question of whether and where in the SG device a spin measurement takes place. The ion interference would also be sensitive to Aharonov-Bohm phase shifts arising from the electromagnetic gauge field. The ion source would be a truly single-particle device [61] and eliminate certain problems arising from particle interactions in high-density sources of neutral bosons [104].

Such single-ion SG devices would open the door for a wide spectrum of fundamental experiments, probing for example weak measurements and Bohmian trajectories. The strong electric interactions may also be used, for example, to entangle the single ion with a solid-state quantum device (an electron in a quantum dot or on a Coulomb island, or a qubit flux gate). This type of interferometer may lead to new sensing capabilities [105]: one of the two ion wavepackets is expected to pass tens to hundreds of nanometers above a surface (in the chip configuration of our proposal [62]) and may probe van der Waals and Casimir-Polder forces, as well as patch potentials. The latter are very important as they are believed to give rise to the anomalous heating observed in miniaturized ion traps [106]. Due to the short distances between the ions and the surface, the device may also be able to sense the gravitational force on small scales [107]. Finally, such a single-ion interferometer may enable searches for exotic physics. These include spontaneous collapse models, the fifth force from a nearby surface, the self-charge interaction between the two ion wavepackets, and so on. Eventually, one may be able to realize a double SG-splitter with different orientations, as originally attempted by Stern, Segrè and co-workers [108, 109], in order to test ideas like the Bohm-Bub non-local hidden variable theory [110–112], or ideas on deterministic quantum mechanics (see, e.g., [113]). Since ions may form the basis of extremely accurate clocks, an ion-SG device would enable clock interferometry at a level sensitive to the Earth's gravitational red-shift (see the proof-of-principle experiments with neutral atoms in [20, 21]). This has important implications for studying the interface between quantum mechanics and general relativity.

7.2   SGI with Massive Objects

The main focus of our future efforts will be to realize an SGI with massive objects. The idea of using the SG interferometer, with a macroscopic object as a probe for gravity, has been detailed in several studies [46, 47, 114, 115] describing a wide range of experiments from the detection of gravitational waves to tests of the quantum nature of gravity. Here we envision using a macroscopic body in the full-loop SGI. We anticipate utilizing spin population oscillations as our interference observable rather than spatial fringes, i.e., density modulations. This observable, as demonstrated in the atomic SGI described above, is advantageous because there is no requirement for long evolution times in order to allow the spatial fringes to develop, nor is high-resolution imaging needed to resolve the spatial fringes. Let us note that there are other proposals to realize a spatial superposition of macroscopic objects [70, 116].

As a specific example, let us consider a solid object comprising $10^6 - 10^{10}$ atoms with a single spin embedded in the solid lattice, e.g., a nano-diamond with a single NV center. Let us first emphasize that even prior to any probing of gravity, a successful SGI will already achieve at least 3 orders of magnitude more atoms than the state-of-the-art in macroscopic-object interferometry [60], thus contributing novel insight to the foundations of quantum mechanics. Another contribution to the foundations of quantum mechanics would be the ability to test continuous spontaneous localization (CSL) models (e.g., [117] and references therein).

When probing gravity, the first contribution of such a massive-object SGI would simply be to measure little g. As the phase is accumulated linearly with the mass, a massive-object interferometer is expected to have much more sensitivity to g than atomic interferometers being used currently (assuming of course that all other features are comparable). This is also a method to verify that a massive-object superposition can be created [114, 118, 119]. A second contribution would measure gravity at short distances, since the massive object may be brought close to a surface while in one of the SGI paths, thus enabling probes of the fifth force. Once the SG technology allows the use of large masses, a third contribution will be the testing of hypotheses concerning gravity self-interaction [48, 116], and once large-area interferometry is also enabled, a fourth contribution would be to detect gravitational waves [46]. Finally, placing two such SGIs in parallel next to each other will enable probes of the quantum nature of gravity [47, 120]. Let us emphasize that, although high accelerations may be obtained with multiple spins, we intend to focus on the case of a macroscopic object with a single spin, since the observable of such a quantum-gravity experiment is entanglement, and averaging over many spins may wash out the signal.

To avoid the hindering consequences of the HD effect, one must ensure that the experimental accuracy of the recombination, as discussed in Sect. 5, will be better than the coherence length. Obviously it is very hard to achieve a large coherence length for a massive object, but recent experimental numbers and estimates seem to indicate that this is feasible. Another crucial problem is the coherence time. A massive object has a huge cross section for interacting with the environment (e.g., background gas), but the extremely short interferometer times, as discussed in this review, seem to serve as a protective shield suppressing decoherence. We are currently a detailed account of these considerations [121].

Acknowledgements We wish to warmly thank all the members—past and present—of the Atom Chip Group at Ben-Gurion University of the Negev, and the team of the BGU nano-fabrication facility for designing and fabricating innovative high-quality chips for our laboratory and for others around the world. The work at BGU described in this review was funded in part by the Israel Science Foundation (1381/13 and 1314/19), the EC "MatterWave" consortium (FP7-ICT-601180), and the German DFG through the DIP program (FO 703/2-1). We also acknowledge support from the PBC program for outstanding postdoctoral researchers of the Israeli Council for Higher Education and from the Ministry of Immigrant Absorption (Israel).

Disclosure Statement The authors declare that they have no competing financial interests.

References

1. W. Gerlach, O. Stern, Der experimentelle Nachweis der Richtungsquantelung im Magnetfeld. Z. Physik **9**, 349 (1922). https://doi.org/10.1007/BF01326983
2. B. Friedrich, D. Herschbach, H. Schmidt-Böcking, J.P. Toennies, An international symposium (Wilhelm and Else Heraeus Seminar #702) marked the centennial of Otto Stern's first molecular beam experiment and the thriving of atomic physics; a European Physical Society Historic Site Was Inaugurated. Front. Phys. **7**, 208 (2019). https://doi.org/10.3389/fphy.2019.00208
3. B. Friedrich, H. Schmidt-Böcking. Otto Stern's molecular beam method and its impact on quantum physics (2020)
4. D. Herschbach, Molecular beams entwined with quantum theory: a bouquet for Max Planck. Ann. Phys. (Leipzig) **10**, 163 (2001). https://doi.org/10.1002/1521-3889(200102)10:1/2<163::AID-ANDP163>3.0.CO;2-W

5. B. Friedrich, D. Herschbach, Stern and Gerlach: how a bad cigar helped reorient atomic physics. Phys. Today **56**, 53 (2003). https://doi.org/10.1063/1.1650229

6. M. Keil, O. Amit, S. Zhou, D. Groswasser, Y. Japha, R. Folman, Fifteen years of cold matter on the atom chip: promise, realizations, and prospects. J. Mod. Opt. **63**, 1840 (2016). https://doi.org/10.1080/09500340.2016.1178820

7. J. Reichel, W. Hänsel, T.W. Hänsch, Atomic micromanipulation with magnetic surface traps. Phys. Rev. Lett. **83**, 3398 (1999). https://doi.org/10.1103/PhysRevLett.83.3398

8. R. Folman, P. Krüger, D. Cassettari, B. Hessmo, T. Maier, J. Schmiedmayer, Controlling cold atoms using nanofabricated surfaces: atom chips. Phys. Rev. Lett. **84**, 4749 (2000). https://doi.org/10.1103/PhysRevLett.84.4749

9. N.H. Dekker, C.S. Lee, V. Lorent, J.H. Thywissen, S.P. Smith, M. Drndić, R.M. Westervelt, M. Prentiss, Guiding neutral atoms on a chip. Phys. Rev. Lett. **84**, 1124 (2000). https://doi.org/10.1103/PhysRevLett.84.1124

10. R. Folman, P. Krüger, J. Schmiedmayer, J. Denschlag, C. Henkel, Microscopic atom optics: from wires to an atom chip. Adv. At. Mol. Opt. Phys. **48**, 263 (2002). https://doi.org/10.1016/S1049-250X(02)80011-8

11. J. Reichel, Microchip traps and Bose-Einstein condensation. Appl. Phys. B **74**, 469 (2002). https://doi.org/10.1007/s003400200861

12. J. Fortágh, C. Zimmermann, Magnetic microtraps for ultracold atoms. Rev. Mod. Phys. **79**, 235 (2007). https://doi.org/10.1103/RevModPhys.79.235

13. *Atom Chips,* J. Reichel, V. Vuletić, eds. (Wiley-VCH, Hoboken, NJ, 2011). https://doi.org/10.1002/9783527633357

14. R. Folman, Material science for quantum computing with atom chips, in *Special Issue on Neutral Particles,*, ed. by R. Folman. Quantum Inf. Process. **10**, 995 (2011), https://doi.org/10.1007/s11128-011-0311-5

15. https://in.bgu.ac.il/en/nano-fab

16. S. Machluf, Y. Japha, R. Folman, Coherent Stern-Gerlach momentum splitting on an atom chip. Nature Commun. **4**, 2424 (2013). https://doi.org/10.1038/ncomms3424

17. Y. Margalit, Z. Zhou, S. Machluf, Y. Japha, S. Moukouri, R. Folman, Analysis of a high-stability Stern-Gerlach spatial fringe interferometer. New J. Phys. **21**, 073040 (2019). https://doi.org/10.1088/1367-2630/ab2fdc

18. Y. Margalit, Z. Zhou, O. Dobkowski, Y. Japha, D. Rohrlich, S. Moukouri, R. Folman, Realization of a complete Stern-Gerlach interferometer. arXiv:1801.02708v2 (2018)

19. O. Amit, Y. Margalit, O. Dobkowski, Z. Zhou, Y. Japha, M. Zimmermann, M.A. Efremov, F.A. Narducci, E.M. Rasel, W.P. Schleich, R. Folman, T^3 Stern-Gerlach matter-wave interferometer. Phys. Rev. Lett. **123**, 083601 (2019). https://doi.org/10.1103/PhysRevLett.123.083601

20. Y. Margalit, Z. Zhou, S. Machluf, D. Rohrlich, Y. Japha, R. Folman, A self-interfering clock as a which-path witness. Science **349**, 1205 (2015). https://doi.org/10.1126/science.aac6498

21. Z. Zhou, Y. Margalit, D. Rohrlich, Y. Japha, R. Folman, Quantum complementarity of clocks in the context of general relativity. Class. Quantum Grav. **35**, 185003 (2018). https://doi.org/10.1088/1361-6382/aad56b

22. A.D. Cronin, J. Schmiedmayer, D.E. Pritchard, Optics and interferometry with atoms and molecules. Rev. Mod. Phys. **81**, 1051 (2009). https://doi.org/10.1103/RevModPhys.81.1051

23. W. Heisenberg. *Die Physikalischen Prinzipien der Quantentheorie.* (S. Hirzel: Leipzig 1930); *The Physical Principles of the Quantum Theory* transl. by C. Eckart, F. C. Hoyt (Dover, Mineola, NY, 1950)

24. D. Bohm, *Quantum Theory* (Prentice-Hall, Englewood Cliffs, 1951), pp. 604–605

25. E.P. Wigner, The problem of measurement. Am. J. Phys. **31**, 6 (1963). https://doi.org/10.1119/1.1969254

26. H. J. Briegel, B.-G. Englert, M.O. Scully, H. Walther, Atom interferometry and the quantum theory of measurement, in *Atom Interferometry*, P.R. Berman, ed. (Academic Press, New York, 1997), p. 240

27. This popular version of the English nursery rhyme appears in https://en.wikipedia.org/wiki/Humpty_Dumpty

28. B.-G. Englert, J. Schwinger, M.O. Scully, Is spin coherence like Humpty-Dumpty? I. Simplified treatment. Found. Phys. **18**, 1045 (1988). https://doi.org/10.1007/BF01909939

29. J. Schwinger, M.O. Scully, B.-G. Englert, Is spin coherence like Humpty-Dumpty? II. General theory. Z. Phys. D **10**, 135 (1988). https://doi.org/10.1007/BF01384847

30. M.O. Scully, B.-G. Englert, J. Schwinger, Spin coherence and Humpty-Dumpty. III. The effects of observation. Phys. Rev. A **40**, 1775 (1989). https://doi.org/10.1103/PhysRevA.40.1775

31. B.-G. Englert, Time reversal symmetry and Humpty-Dumpty. Z. Naturforsch A **52**, 13 (1997). https://doi.org/10.1515/zna-1997-1-206

32. M. O. Scully, Jr. W.E. Lamb, A. Barut, On the theory of the Stern-Gerlach apparatus. Found. Phys. **17**, 575 (1987).https://doi.org/10.1007/BF01882788

33. T.R. de Oliveira, A.O. Caldeira, Dissipative Stern-Gerlach recombination experiment. Phys. Rev. A **73**, 042502 (2006). https://doi.org/10.1103/PhysRevA.73.042502

34. M. Devereux, Reduction of the atomic wavefunction in the Stern-Gerlach magnetic field. Can. J. Phys. **93**, 1382 (2015). https://doi.org/10.1139/cjp-2015-0031

35. J. Robert, C. Miniatura, S. Le Boiteux, J. Reinhardt, V. Bocvarski, J. Baudon, Atomic interferometry with metastable hydrogen atoms. Europhys. Lett. **16**, 29 (1991). https://doi.org/10.1209/0295-5075/16/1/006

36. C. Miniatura, F. Perales, G. Vassilev, J. Reinhardt, J. Robert, J. Baudon, A longitudinal Stern-Gerlach interferometer: the "beaded" atom. J. Phys. II **1**, 425 (1991). https://doi.org/10.1051/jp2:1991177

37. C. Miniatura, J. Robert, S. Le Boiteux, J. Reinhardt, J. Baudon, A longitudinal Stern-Gerlach atomic interferometer. App. Phys. B **54**, 347 (1992). https://doi.org/10.1007/BF00325378

38. J. Robert, C. Miniatura, O. Gorceix, S. Le Boiteux, V. Lorent, J. Reinhardt, J. Baudon, Atomic quantum phase studies with a longitudinal Stern-Gerlach interferometer. J. Phys. II **11**, 601 (1992). https://doi.org/10.1051/jp2:1992155

39. C. Miniatura, J. Robert, O. Gorceix, V. Lorent, S. Le Boiteux, J. Reinhardt, J. Baudon, Atomic interferences and the topological phase. Phys. Rev. Lett. **69**, 261 (1992). https://doi.org/10.1103/PhysRevLett.69.261

40. S. Nic Chormaic, V. Wiedemann, C. Miniatura, J. Robert, S. Le Boiteux, V. Lorent, O. Gorceix, S. Feron, J. Reinhardt, J. Baudon, Longitudinal Stern-Gerlach atomic interferometry using velocity selected atomic beams. J. Phys. B **26**, 1271 (1993). https://doi.org/10.1088/0953-4075/26/7/011

41. J. Baudon, R. Mathevet, J. Robert, Atomic interferometry. J. Phys. B **32**, R173 (1999). https://doi.org/10.1088/0953-4075/32/15/201

42. M. Boustimi, V. Bocvarski, K. Brodsky, F. Perales, J. Baudon, J. Robert, Atomic interference patterns in the transverse plane. Phys. Rev. A **61**, 033602 (2000). https://doi.org/10.1103/PhysRevA.61.033602

43. E. Maréchal, R. Long, T. Miossec, J.-L. Bossennec, R. Barbé, J.-C. Keller, O. Gorceix, Atomic spatial coherence monitoring and engineering with magnetic fields. Phys. Rev. A **62**, 53603 (2000). https://doi.org/10.1103/PhysRevA.62.053603

44. B. Viaris de Lesegno, J.C. Karam, M. Boustimi, F. Perales, C. Mainos, J. Reinhardt, J. Baudon, V. Bocvarski, D. Grancharova, F. Pereira Dos Santos, T. Durt, H. Haberland J. Robert, Stern Gerlach interferometry with metastable argon atoms: an immaterial mask modulating the profile of a supersonic beam. Eur. Phys. J. D **23**,25 (2003). https://doi.org/10.1140/epjd/e2003-00023-y

45. K. Rubin, M. Eminyan, F. Perales, R. Mathevet, K. Brodsky, B. Viaris de Lesegno, J. Reinhardt, M. Boustimi, J. Baudon, J.-C. Karam, J. Robert, Atom interferometer using two Stern-Gerlach magnets. Laser Phys. Lett. **1**, 184 (2004). https://doi.org/10.1002/lapl.200310047

46. R.J. Marshman, A. Mazumdar, G.W. Morley, P.F. Barker, S. Hoekstra, S. Bose, Mesoscopic interference for metric and curvature (MIMAC) & gravitational wave detection. New J. Phys. **22**, 083012 (2020). https://doi.org/10.1088/1367-2630/ab9f6c

47. S. Bose, A. Mazumdar, G.W. Morley, H. Ulbricht, M. Toroš, M. Paternostro, A.A. Geraci, P.F. Barker, M.S. Kim, G. Milburn, Spin entanglement witness for quantum gravity. Phys. Rev. Lett. **119**, 240401 (2017). https://doi.org/10.1103/PhysRevLett.119.240401

48. M. Hatifi, T. Durt, Revealing self-gravity in a Stern-Gerlach Humpty-Dumpty experiment. arXiv:2006.07420 (2020)
49. C. Marletto, V. Vedral, On the testability of the equivalence principle as a gauge principle detecting the gravitational t^3 phase. Front. Phys. **8**, 176 (2020). https://doi.org/10.3389/fphy. 2020.00176
50. M. Gebbe, S. Abend, J.-N. Siemß, M. Gersemann, H. Ahlers, H.Müntinga, S. Herrmann, N. Gaaloul, C. Schubert, K. Hammerer, C. Lämmerzahl, W. Ertmer, E.M. Rasel, Twin-lattice atom interferometry. arXiv:1907.08416v1 (2019)
51. B. Canuel, S. Abend, P. Amaro-Seoane, F. Badaracco, Q. Beaufils, A. Bertoldi, K. Bongs, P. Bouyer, C. Braxmaier, W. Chaibi, N. Christensen, F. Fitzek, G. Flouris, N. Gaaloul, S. Gaffet, C.L. Garrido Alzar, R. Geiger, S. Guellati-Khelifa, K. Hammerer, J. Harms, J. Hinderer, M. Holynski, J. Junca, S. Katsanevas, C. Klempt, C. Kozanitis, M. Krutzik, A. Landragin, I. Làzaro Roche, B. Leykauf, Y.-H. Lien, S. Loriani, S. Merlet, M. Merzougui, M. Nofrarias, P. Papadakos, F. Pereira dos Santos, A. Peters, D. Plexousakis, M. Prevedelli, E.M. Rasel, Y. Rogister, S. Rosat, A. Roura, D. O. Sabulsky, V. Schkolnik, D. Schlippert, C. Schubert, L. Sidorenkov, J.-N. Siemß, C. F. Sopuerta, F. Sorrentino, C. Struckmann, G.M. Tino, G. Tsagkatakis, A. Viceré, W. von Klitzing, L. Woerner, X. Zou, Technologies for the ELGAR large scale atom interferometer array. arXiv:2007.04014v1 (2020); ELGAR–a European Laboratory for Gravitation and Atom-interferometric Research. Class. Quantum Grav. **37**, 225017 (2020). https://doi.org/10.1088/1361-6382/aba80e
52. J. Rudolph, T. Wilkason, M. Nantel, H. Swan, C.M. Holland, Y. Jiang, B.E. Garber, S.P. Carman, J.M. Hogan, Large momentum transfer clock atom interferometry on the 689 nm intercombination line of strontium. Phys. Rev. Lett. **124**, 083604 (2020). https://doi.org/10. 1103/PhysRevLett.124.083604
53. D.V. Strekalov, N. Yu, K. Mansour, *Sub-shot Noise Power Source for Microelectronics.* NASA Tech Briefs (Pasadena, CA, 2011). http://ntrs.nasa.gov/archive/nasa/casi.ntrs.nasa. gov/20120006513.pdf
54. E. Danieli, J. Perlo, B. Blümich, F. Casanova, Highly stable and finely tuned magnetic fields generated by permanent magnet assemblies. Phys. Rev. Lett. **110**, 180801 (2013). https://doi. org/10.1103/PhysRevLett.110.180801
55. S. Zhou, D. Groswasser, M. Keil, Y. Japha, R. Folman, Robust spatial coherence 5 μm from a room-temperature atom chip. Phys. Rev. A **93**, 063615 (2016). https://doi.org/10.1103/ PhysRevA.93.063615
56. P. Treutlein, P. Hommelhoff, T. Steinmetz, T.W. Hänsch, J. Reichel, Coherence in microchip traps. Phys. Rev. Lett. **92**, 203005 (2004). https://doi.org/10.1103/PhysRevLett.92.203005
57. F. Cerisola, Y. Margalit, S. Machluf, A.J. Roncaglia, J.P. Paz, R. Folman, Using a quantum work meter to test non-equilibrium fluctuation theorems. Nature Commun. **8**, 1241 (2017). https://doi.org/10.1038/s41467-017-01308-7
58. B.S. Zhao, W. Zhang, W. Schöllkopf, Non-destructive quantum reflection of helium dimers and trimers from a plane ruled grating. Mol. Phys. **111**, 1772 (2013). https://doi.org/10.1080/ 00268976.2013.787150
59. S. Zeller, M. Kunitski, J. Voigtsberger, A. Kalinin, A. Schottelius, C. Schober, M. Waitz, H. Sann, A. Hartung, T. Bauer, M. Pitzer, F. Trinter, C. Goihl, C. Janke, M. Richter, G. Kastirke, M. Weller, A. Czasch, M. Kitzler, M. Braune, R.E. Grisenti, W. Schöllkopf, L.P.H. Schmidt, M.S. Schöffler, J.B. Williams, T. Jahnke, R.Dörner, Imaging the He$_2$ quantum halo state using a free electron laser. Proc. Natl. Acad. Sci. USA **113**, 14651 (2016). https://doi.org/10.1073/ pnas.1610688113
60. Y.Y. Fein, P. Geyer, P. Zwick, F. Kiałka, S. Pedalino, M. Mayor, S. Gerlich, M. Arndt, Quantum superposition of molecules beyond 25 kDa. Nature Phys. **15**, 1242 (2019). https://doi.org/10. 1038/s41567-019-0663-9
61. G. Jacob, K. Groot-Berning, S. Wolf, S. Ulm, L. Couturier, S.T. Dawkins, U.G. Poschinger, F. Schmidt-Kaler, K. Singer, Transmission microscopy with nanometer resolution using a deterministic single ion source. Phys. Rev. Lett. **117**, 043001 (2016). https://doi.org/10.1103/ PhysRevLett.117.043001

62. C. Henkel, G. Jacob, F. Stopp, F. Schmidt-Kaler, M. Keil, Y. Japha, R. Folman, Stern-Gerlach splitting of low-energy ion beams. New J. Phys. **21**, 083022 (2019). https://doi.org/10.1088/1367-2630/ab36c7

63. L.W. Bruch, W. Schöllkopf, J.P. Toennies, The formation of dimers and trimers in free jet ^{4}He cryogenic expansions. J. Chem. Phys. **117**, 1544 (2002). https://doi.org/10.1063/1.1486442

64. Y. Margalit, *Stern-Gerlach Interferometry with Ultracold Atoms*. (Ph.D. Thesis, Ben-Gurion University, 2018). http://www.bgu.ac.il/atomchip/Theses/Yair_Margalit_PhD_Thesis_2018.pdf

65. F. Kiałka, B.A. Stickler, K. Hornberger, Y.Y. Fein, P. Geyer, L. Mairhofer, S. Gerlich, M. Arndt, Concepts for long-baseline high-mass matter-wave interferometry. Phys. Scripta **94**, 034001 (2019). https://doi.org/10.1088/1402-4896/aaf243

66. D.A. Steck. Rubidium 87 D Line Data (2003). https://steck.us/alkalidata/rubidium87numbers.1.6.pdf

67. S. Machluf, *Coherent Splitting of Matter-Waves on an Atom Chip Using a State-Dependent Magnetic Potential* (Ph.D. Thesis, Ben-Gurion University, 2013). http://www.bgu.ac.il/atomchip/Theses/Shimon_Machluf_PhD_2013.pdf

68. O. Amit, *Matter-Wave Interferometry on an Atom Chip* (Ph.D. Thesis, Ben-Gurion University, 2020). http://www.bgu.ac.il/atomchip/Theses/PhD_Thesis_Omer_Amit_submitted.pdf

69. D.E. Miller, J.R. Anglin, J.R. Abo-Shaeer, K. Xu, J.K. Chin, W. Ketterle, High-contrast interference in a thermal cloud of atoms. Phys. Rev. A **71**, 043615 (2005). https://doi.org/10.1103/PhysRevA.71.043615

70. O. Romero-Isart, Coherent inflation for large quantum superpositions of levitated microspheres. New J. Phys. **19**, 719711 (2017). https://doi.org/10.1088/1367-2630/aa99bf

71. R. Colella, A.W. Overhauser, S.A. Werner, Observation of gravitationally induced quantum interference. Phys. Rev. Lett. **34**, 1472 (1975). https://doi.org/10.1103/PhysRevLett.34.1472

72. M. Zych, F. Costa, I. Pikovski, Č. Brukner, Quantum interferometric visibility as a witness of general relativistic proper time. Nature Commun. **2**, 505 (2011). https://doi.org/10.1038/ncomms1498

73. S. Dimopoulos, P.W. Graham, J.M. Hogan, M.A. Kasevich, Testing general relativity with atom interferometry. Phys. Rev. Lett. **98**, 111102 (2007). https://doi.org/10.1103/PhysRevLett.98.111102

74. H. Müntinga, H. Ahlers, M. Krutzik, A. Wenzlawski, S. Arnold, D. Becker, K. Bongs, H. Dittus, H. Duncker, N. Gaaloul, C. Gherasim, E. Giese, C. Grzeschik, T.W. Hänsch, O. Hellmig, W. Herr, S. Herrmann, E. Kajari, S. Kleinert, C. Lämmerzahl, W. Lewoczko-Adamczyk, J. Malcolm, N. Meyer, R. Nolte, A. Peters, M. Popp, J. Reichel, A. Roura, J. Rudolph, M. Schiemangk, M. Schneider, S.T. Seidel, K. Sengstock, V. Tamma, T. Valenzuela, A. Vogel, R. Walser, T. Wendrich, P. Windpassinger, W. Zeller, T. van Zoest, W. Ertmer, W.P. Schleich, E.M. Rasel, Interferometry with Bose-Einstein condensates in microgravity. Phys. Rev. Lett. **110**, 093602 (2013). https://doi.org/10.1103/PhysRevLett.110.093602

75. C.C.N. Kuhn, G.D. McDonald, K.S. Hardman, S. Bennetts, P.J. Everitt, P.A. Altin, J.E. Debs, J.D. Close, N.P. Robins, A. Bose-condensed, simultaneous dual-species Mach-Zehnder atom interferometer. New J. Phys. **16**, 073035 (2014). https://doi.org/10.1088/1367-2630/16/7/073035

76. I. Pikovski, M. Zych, F. Costa, Č. Brukner, Universal decoherence due to gravitational time dilation. Nat. Phys. **11**, 668 (2015). https://doi.org/10.1038/nphys3366

77. B.-G. Englert, Fringe visibility and which-way information: an inequality. Phys. Rev. Lett. **77**, 2154 (1996). https://doi.org/10.1103/PhysRevLett.77.2154

78. E. Giese, A. Friedrich, F. Di Pumpo, A. Roura, W.P. Schleich, D.M. Greenberger, E.M. Rasel, Proper time in atom interferometers: diffractive versus specular mirrors. Phys. Rev. Lett. **99**, 013627 (2019). https://doi.org/10.1103/PhysRevA.99.013627

79. M. Lugli, *Mass and Proper Time as Conjugated Observables* (Graduate Thesis, Università degli Studi di Pavia, Italy, 2017). http://arxiv.org/abs/1710.06504

80. J. Samuel, R. Bhandari, General setting for Berry's phase. Phys. Rev. Lett. **60**, 2339 (1988). https://doi.org/10.1103/PhysRevLett.60.2339

81. R. Bhandari, SU(2) phase jumps and geometric phases. Phys. Lett. A **157**, 221 (1991). https://doi.org/10.1016/0375-9601(91)90055-D

82. T. van Dijk, H.F. Schouten, T.D. Visser, Geometric interpretation of the Pancharatnam connection and non-cyclic polarization changes. J. Am. Opt. Soc. A **27**, 1972 (2010). https://doi.org/10.1364/JOSAA.27.001972

83. S. Pancharatnam, Generalized theory of interference, and its applications. Proc. Indian Acad. Sci. A **44**, 247 (1956). https://doi.org/10.1007/BF03046050

84. Z. Zhou, Y. Margalit, S. Moukouri, Y. Meir, R. Folman, An experimental test of the geodesic rule proposition for the noncyclic geometric phase. Sci. Adv. **6**, eaay8345 (2020). https://doi.org/10.1126/sciadv.aay8345

85. E.H. Kennard, Zur Quantenmechanik einfacher Bewegungstypen. Z. fur Physik **44**, 326 (1927). https://doi.org/10.1007/BF01391200

86. E.H. Kennard, The quantum mechanics of an electron or other particle. J. Franklin Inst. **207**, 47 (1929). https://doi.org/10.1016/S0016-0032(29)91274-6

87. G.G. Rozenman, M. Zimmermann, M.A. Efremov, W.P. Schleich, L. Shemer, A. Arie, Amplitude and phase of wave packets in a linear potential. Phys. Rev. Lett. **122**, 124302 (2019). https://doi.org/10.1103/PhysRevLett.122.124302

88. G.G. Rozenman, L. Shemer, A. Arie, Observation of accelerating solitary wavepackets. Phys. Rev. E **101**, 050201(R) (2020). https://doi.org/10.1103/PhysRevE.101.050201

89. G.G. Rozenman, M. Zimmermann, M.A. Efremov, W.P. Schleich, W.B. Case, D.M. Greenberger, L. Shemer, A. Arie, Projectile motion of surface gravity water wave packets: An analogy to quantum mechanics. Eur. Phys. J. Spec. Top. (2021). https://doi.org/10.1140/epjs/s11734-021-00096-y

90. M. Zimmermann, M. Efremov, W. Zeller, W. Schleich, J. Davis, F. Narducci, Representation-free description of atom interferometers in time-dependent linear potentials. New J. Phys. **21**, 073031 (2019). https://doi.org/10.1088/1367-2630/ab2e8c

91. M. Efremov, M. Zimmermann, O. Amit, Y. Margalit, R. Folman, W. Schleich, Atomic interferometer sensitive to time-dependent acceleration. In preparation (2021)

92. C.J. Bordé, Atomic interferometry with internal state labelling. Phys. Lett. A **140**, 10 (1989). https://doi.org/10.1016/0375-9601(89)90537-9

93. M. Kasevich, S. Chu, Atomic interferometry using stimulated Raman transitions. Phys. Rev. Lett. **67**, 181 (1991). https://doi.org/10.1103/PhysRevLett.67.181

94. A. Peters, K. Chung, S. Chu, High-precision gravity measurements using atom interferometry. Metrologia **38**, 25 (2001). https://doi.org/10.1088/0026-1394/38/1/4

95. L. Brillouin, Is it possible to test by a direct experiment the hypothesis of the spinning electron? Proc. Natl. Acad. Sci. **14**, 755 (1928). https://doi.org/10.1073/pnas.14.10.755

96. N. Bohr, Chemistry and the quantum theory of atomic constitution. J. Chem. Soc. **349**, (1932). https://doi.org/10.1039/JR9320000349

97. G.A. Gallup, H. Batelaan, T.J. Gay, Quantum-mechanical analysis of a longitudinal Stern-Gerlach effect. Phys. Rev. Lett. **86**, 4508 (2001). https://doi.org/10.1103/PhysRevLett.86.4508

98. H. Batelaan, T.J. Gay, J.J. Schwendiman, Stern-Gerlach effect for electron beams. Phys. Rev. Lett. **79**, 4517 (1997). https://doi.org/10.1103/PhysRevLett.79.4517

99. H. Batelaan, Electrons, Stern-Gerlach magnets, and quantum mechanical propagation. Am. J. Phys. **70**, 325 (2002). https://doi.org/10.1119/1.1450559

100. B.M. Garraway, S. Stenholm, Does a flying electron spin? Contemp. Phys. **43**, 147 (2002). https://doi.org/10.1080/00107510110102119

101. G. Jacob, *Ion Implantation and Transmission Microscopy with Nanometer Resolution Using a Deterministic Ion Source*. (Ph.D. Dissertation, Johannes-Gutenberg-Universität, Mainz, Germany, 2016). https://www.quantenbit.physik.uni-mainz.de/files/2019/10/DissSW.pdf

102. E.A. Hinds, I.G. Hughes, Magnetic atom optics: Mirrors, guides, traps, and chips for atoms. J. Phys. D **32**, R119–R146 (1999). http://iopscience.iop.org/article/10.1088/0022-3727/32/18/201/pdf

103. T.D. Tran, Y. Wang, A. Glaetzle, S. Whitlock, A. Sidorov, P. Hannaford, Magnetic lattices for ultracold atoms (2019). arXiv:1906.08918

104. G.M. Tino, M.A. Kasevich, eds., Atom iterferometry, in *Proceedings of International School of Physics "Enrico Fermi"*. http://ebooks.iospress.nl/volume/atom-interferometry, Vol 188 (IOS Press, Amsterdam, 2014)
105. F. Hasselbach, Progress in electron- and ion-interferometry. Rep. Prog. Phys. **73**, 016101 (2010). https://doi.org/10.1088/0034-4885/73/1/016101
106. M. Brownnutt, M. Kumph, P. Rabl, R. Blatt, Ion-trap measurements of electric-field noise near surfaces. Rev. Mod. Phys. **87**, 1419 (2015). https://doi.org/10.1103/RevModPhys.87. 1419
107. R.O. Behunin, D.A.R. Dalvit, R.S. Decca, C.C. Speake, Limits on the accuracy of force sensing at short separations due to patch potentials. Phys. Rev. D **89**, 051301(R) (2014). https://doi.org/10.1103/PhysRevD.89.051301
108. T.E. Phipps, O. Stern, Über die Einstellung der Richtungsquantelung. Z. Phys. **73**, 185 (1932). https://doi.org/10.1007/BF01351212
109. R. Frisch, E. Segrè, Über die Einstellung der Richtungsquantelung II. Z. Phys. **80**, 610 (1933). https://doi.org/10.1007/BF01335699
110. D. Bohm, J. Bub, A proposed solution of the measurement problem in quantum mechanics by a hidden variable theory. Rev. Mod. Phys. **38**, 453 (1966). https://doi.org/10.1103/ RevModPhys.38.453
111. R. Folman, A search for hidden variables in the domain of high energy physics. Found. Phys. Lett. **7**, 191 (1994). https://doi.org/10.1007/BF02415510
112. S. Das, M. Nöth, D. Dürr, Exotic Bohmian arrival times of spin-1/2 particles: an analytical treatment. Phys. Rev. A **99**, 052124 (2019). https://doi.org/10.1103/PhysRevA.99.052124
113. L.S. Schulman, Program for the special state theory of quantum measurement. Entropy **19**, 343 (2017). https://doi.org/10.3390/e19070343
114. C. Wan, M. Scala, G.W. Morley, A.A.T.M. Rahman, H. Ulbricht, J. Bateman, P.F. Barker, S. Bose, M.S. Kim, Free nano-object Ramsey interferometry for large quantum superpositions. Phys. Rev. Lett. **117**, 143003 (2016). https://doi.org/10.1103/PhysRevLett.117.143003
115. R.J. Marshman, A. Mazumdar, S. Bose, Locality and entanglement in table-top testing of the quantum nature of linearized gravity. Phys. Rev. A **101**, 052110 (2020). https://doi.org/10. 1103/PhysRevA.101.052110
116. H. Pino, J. Prat-Camps, K. Sinha, B.P. Venkatesh, O. Romero-Isart, On-chip quantum interference of a superconducting microsphere. Quantum Sci. Technol. **3**, 25001 (2018). https:// doi.org/10.1088/2058-9565/aa9d15
117. O. Romero-Isart, Quantum superposition of massive objects and collapse models. Phys. Rev. A **84**, 052121 (2011). https://doi.org/10.1103/PhysRevA.84.052121
118. M. Scala, M.S. Kim, G.W. Morley, P.F. Barker, S. Bose, Matter-wave interferometry of a levitated thermal nano-oscillator induced and probed by a spin. Phys. Rev. Lett. **111**, 180403 (2013). https://doi.org/10.1103/PhysRevLett.111.180403
119. M. Toroš, S. Bose, P.F. Barker, Atom-Nanoparticle Schrödinger Cats. arXiv:2005.12006 (2020)
120. C. Marletto, V. Vedral, Gravitationally induced entanglement between two massive particles is sufficient evidence of quantum effects in gravity. Phys. Rev. Lett. **119**, 240402 (2017). https://doi.org/10.1103/PhysRevLett.119.240402
121. Y. Margalit, O. Dobkowski, Z. Zhou, O. Amit, Y. Japha, S. Moukouri, D. Rohrlich, A. Mazumdar, S. Bose, C. Henkel, R. Folman, Realization of a complete Stern-Gerlach interferometer: towards a test of quantum gravity (2020). https://arxiv.org/pdf/2011.10928v1.pdf

From Hot Beams to Trapped Ultracold Molecules: Motivations, Methods and Future Directions

N. J. Fitch and M. R. Tarbutt

Abstract Over the past century, the molecular beam methods pioneered by Otto Stern have advanced our knowledge and understanding of the world enormously. Stern and his colleagues used these new techniques to measure the magnetic dipole moments of fundamental particles with results that challenged the prevailing ideas in fundamental physics at that time. Similarly, recent measurements of fundamental electric dipole moments challenge our present day theories of what lies beyond the Standard Model of particle physics. Measurements of the electron's electric dipole moment (eEDM) rely on the techniques invented by Stern and later developed by Rabi and Ramsey. We give a brief review of this historical development and the current status of eEDM measurements. These experiments, and many others, are likely to benefit from ultracold molecules produced by laser cooling. We explain how laser cooling can be applied to molecules, review recent progress in this field, and outline some eagerly anticipated applications.

1 Introduction

It has been nearly a hundred years since Otto Stern and Walther Gerlach used an atomic beam to reveal one of the most striking aspects of the then-burgeoning quantum theory, space quantization [1]. Their work introduced new techniques that would later be used in countless experiments in physics and chemistry. Stern saw clearly the great promise of his new method, stating [2]

> The molecular beam method must be made so sensitive that in many instances it will become possible to measure effects and tackle new problems which presently are not accessible with known experimental methods.

He was right. The molecular beam method has been at the heart of atomic and molecular physics ever since and remains the method of choice for a huge number of experiments.

N. J. Fitch · M. R. Tarbutt (✉)
Centre for Cold Matter, Blackett Laboratory, Imperial College London, Prince Consort Road, London SW7 2AZ, UK
e-mail: m.tarbutt@imperial.ac.uk

A more recent development—laser cooling—can also be traced back to Stern, via Frisch who used the molecular beam method to measure the photon recoil momentum [3]. By controlling this recoil, modern atomic physics experiments routinely cool atoms and ions to μK temperatures. Until recently, experiments with molecules lagged behind, usually because of the difficulty of cooling them. Nevertheless, molecules have many useful properties that are increasingly being exploited for a variety of applications including tests of fundamental physics. An important example is the measurement of the electron's electric dipole moment (eEDM) where the precision of molecular experiments exceeds that achieved using atoms. The desire to improve these measurements provides strong motivation to extend cooling and trapping methods to molecules, and this has been achieved in the last decade. Laser cooling has been used to collimate and decelerate molecular beams, capture molecules in magneto-optical traps, and then cool them to ultracold temperature. These ultracold molecules can be used to address a wide variety of important problems - exploring what new forces lie beyond the Standard Model of particle physics [4, 5], studying collisions and reactions at the quantum level [6, 7], simulating the behaviour of many-body quantum systems [8–10], and processing quantum information [11, 12].

In this article, we review some of these past developments and future prospects. We begin in Sect. 2 with a brief review of molecular beam sources. In Sect. 3 we consider how the development of molecular beam methods for measuring magnetic dipole moments eventually enabled measurements of the electric dipole moments of fundamental particles. We briefly review the current status of these experiments in Sect. 4 and explain the importance of laser cooling to the future of this endeavour. In Sects. 5 and 6 we explain how laser cooling works for molecules and present recent achievements in this field. Finally, in Sect. 7, we give a brief overview of some applications of ultracold molecules and how they might be realized.

2 Molecular Beam Sources

The molecular beam method developed by Stern [13, 14] is the foundation for innumerable experiments in atomic and molecular physics. Here, we give a brief review of the three main types of atomic and molecular beam sources in use today: effusive beams, supersonic beams, and cryogenic buffer gas beams. Their velocity distributions and flux are compared in Fig. 1 for the case of YbF molecules, one of the few species for which all three types of beam source have been realized.

Effusive sources typically use heated ovens to generate a sufficient vapour pressure of the atoms or molecules of interest. They operate at low pressure so that there are no collisions in the vicinity of the exit aperture. As first shown experimentally by Stern [13, 14], these sources produce beams with a broad velocity distribution, whose mean and width both scale as $\sqrt{T/m_s}$, where T is the temperature of the source and m_s is the mass of the species. As a consequence of the high oven temperature, effusive beams are characterized by a wide velocity distribution and low flux in any single quantum state.

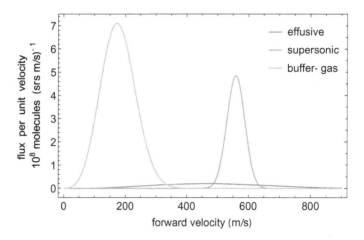

Fig. 1 Velocity distributions of effusive, supersonic, and buffer-gas beams. The vertical axis indicates the number of molecules per unit solid angle per unit time and per unit interval of velocity. The plots are for YbF molecules in their absolute ground state, chosen because, for this molecule, all three sources have been developed and characterised. Realistic operating conditions have been taken. For the effusive case, molecules are generated from an oven at 1500 K [15]. For the supersonic case, a carrier gas of argon with a reservoir temperature of 300 K is assumed, with an internal beam temperature of 4 K and a repetition rate of 25 Hz [16]. For the cryogenic buffer-gas case, a carrier gas of helium at 4 K creates a beam with moderate hydrodynamic boosting, operating at 10 Hz [17]

In supersonic sources [18, 19], a gas held at high pressure expands through a nozzle into a vacuum chamber. There are a large number of collisions in the vicinity of the nozzle. The slower particles are bumped from behind, while the faster ones bump into those ahead, so that all the particles end up travelling at nearly the same speed. In this way, the random thermal motion is converted into forward kinetic energy, producing a cold, fast beam—the mean velocity is high, but the velocity distribution is narrow, as illustrated in Fig. 1. The collisions also transfer the rotational and vibrational energy of the molecules to forward kinetic energy, resulting in a beam that is cold in all degrees of freedom. The first supersonic beams were continuous, but the method was soon extended by using pulsed valves with short opening times. In this way, intense pulses can be produced without an excessive gas load. A wide variety of methods have been developed to introduce atoms and molecules of interest into the supersonic expansion, including laser ablation, electric discharge, and photodissociation. Translational temperatures of 1 K are typical, and beam speeds are 1800 m/s when the carrier gas is room temperature helium, and 400 m/s for room temperature krypton.

The third type of molecular beam source is the cryogenic buffer gas source [20, 21]. Here, the molecules of interest are formed inside a cryogenically-cooled cell containing a cold buffer gas, often helium at 4 K. The molecules are commonly formed by laser ablation or introduced into the cell through a capillary. The internal and motional degrees of freedom of the molecules thermalize through collisions with the buffer gas, and then the molecules exit through a hole in the cell to form a beam. The density of buffer gas in the cell is determined by the gas flow rate and the size of the exit hole. When the density is low, there are few collisions near the aperture, so the beam tends towards the effusive regime. These beams are slow, especially for heavy species, for then the mass m_s is large whereas the temperature T is small, typically two orders of magnitude lower than a standard effusive source. Beam speeds as low as 40 m/s have been achieved this way [22]. However, the molecular flux tends to be low in this regime because most molecules diffuse to the cell walls, where they freeze, instead of passing through the exit aperture. As the buffer gas flow is increased it sweeps more molecules out of the cell, increasing the beam flux. However, collisions near the aperture boost the beam speed. In the limit of high density the speed of the molecules reaches the supersonic speed of the buffer gas which scales as $\sqrt{T/m_b}$, where m_b is the mass of the buffer gas atoms. Very often, cryogenic buffer gas sources are operated in an intermediate flow regime where the flow is high enough to extract a substantial fraction of the molecules from the cell, but low enough for a moderate beam speed. Speeds in the range 100–200 m/s are typical, as illustrated in Fig. 1. Due to the high flux and low relative beam speeds, these sources are becoming increasingly popular, especially for experiments on laser cooling of molecules and tests of fundamental physics.

3 Particle Dipole Moments

Stern's pioneering experiments established the reality of space quantization and determined the magnetic dipole moments of the electron and proton [1, 23–25]. It is interesting to consider whether elementary particles might also have *electric* dipole moments. Just like the magnetic moment, such an electric dipole would have to be oriented along the particle's spin. Furthermore, this orientation must be fixed, since the particle would otherwise have an additional degree of freedom that would, for example, change the filling of electron orbitals in the periodic table. A spin defines a direction of circulation, as does a magnetic moment, so it seems natural for the two to be associated. Far less natural is the association of an electric dipole – a charge separation – with this direction of circulation. Indeed, such an electric dipole moment (EDM) implies a difference between left- and right-handed coordinate systems, and implies a fundamental arrow of time. To see this, consider the Hamiltonian for a particle with magnetic moment $\boldsymbol{\mu}$ and EDM $\boldsymbol{d}$, both fixed relative to the spin, interacting with magnetic and electric fields $\boldsymbol{B}$ and $\boldsymbol{E}$:

$$\mathcal{H} = -\boldsymbol{\mu} \cdot \boldsymbol{B} - \boldsymbol{d} \cdot \boldsymbol{E}. \qquad (1)$$

Reflection in a mirror, equivalent to the parity operation, reverses $\boldsymbol{E}$ but does not reverse $\boldsymbol{B}$, $\boldsymbol{\mu}$ or $\boldsymbol{d}$. Conversely, reversing the direction of time reverses $\boldsymbol{B}$, $\boldsymbol{\mu}$ and $\boldsymbol{d}$, but not $\boldsymbol{E}$. We see that while the first term in $\mathcal{H}$ is even under both the parity and time-reversal operations, the second term is odd and so the existence of an EDM violates both symmetries.

Prior to the 1950s, it was generally considered that nature did not distinguish between left and right, or between forwards and backwards in time, and this seemed to be a powerful argument against the existence of fundamental electric dipoles, implying that $|\boldsymbol{d}| = 0$. This idea was challenged by Purcell and Ramsey who insisted that it was "*a purely experimental matter*", noted that existing evidence was weak, and declared their intention to measure the EDM of the neutron [26]. Regarding the need for experimental evidence to determine whether parity (P), time-reversal (T) and charge conjugation (C) are symmetries of nature, Hermann Weyl was similarly emphatic, writing that [27]

> a priori evidence is not sufficient to settle the question; the empirical facts have to be consulted.

Along this line of thought, in 1956 Lee and Yang [28] noted that, for the weak interaction,

> parity conservation is so far only an extrapolated hypothesis unsupported by experimental evidence.

Within a few months it was discovered that the weak interaction violates P symmetry [29–31]. In 1964 it was found that the weak interaction also violates CP symmetry, the combined symmetry of charge conjugation and parity [32]. CP violation is equivalent to T-violation in most theories, and so the last theoretical objection to the existence of fundamental EDMs was removed. Suddenly, and ever since, the question was not whether particles could have electric dipoles, but why those dipoles are so small.

In considering how the electric dipole moment of a particle might be measured, it is instructive to reflect on Stern's method for measuring magnetic moments, which is illustrated in Fig. 2a. The magnetic moment of an atom is proportional to its internal angular momentum, $\hbar \boldsymbol{F}$, so we often write $\boldsymbol{\mu} = -g \mu_B \boldsymbol{F}$, where μ_B is the Bohr magneton and g is a proportionality factor. Taking $\boldsymbol{B}$ in the z-direction, the interaction energy is $W_B = \langle -\boldsymbol{\mu} \cdot \boldsymbol{B} \rangle = g \mu_B B \langle F_z \rangle = g \mu_B B \, m_F$, where m_F is the projection of the angular momentum onto the z-axis. Space quantization is expressed by only discrete values being allowed for m_F. In the inhomogeneous field of the Stern-Gerlach magnets there is a force on the atoms $\boldsymbol{F}_B = -\nabla W_B = -g \mu_B \nabla B \, m_F$, leading to the deflection proportional to m_F observed by Stern and Gerlach. Let us now consider an atom with an unpaired electron that has an eEDM, d_e. If an electric field E is applied in the z-direction, there is an interaction energy $W_E = -R d_e E \, m_F / |F|$, analogous to the magnetic interaction. The proportionality factor R depends on the choice of atom and is often called the enhancement factor because, for heavy atoms, $|R|$ can be considerably larger than 1 [33]. For example, for Cs, $R = 120$ [34]. Suppose we pass a beam of Cs atoms through a region of inhomogeneous electric field. Of course, the electric field produces an induced electric dipole in the atom, resulting in a force which is often used to deflect atoms and molecules. This force is the same for states of opposite m_F and is not the one of interest here. The force

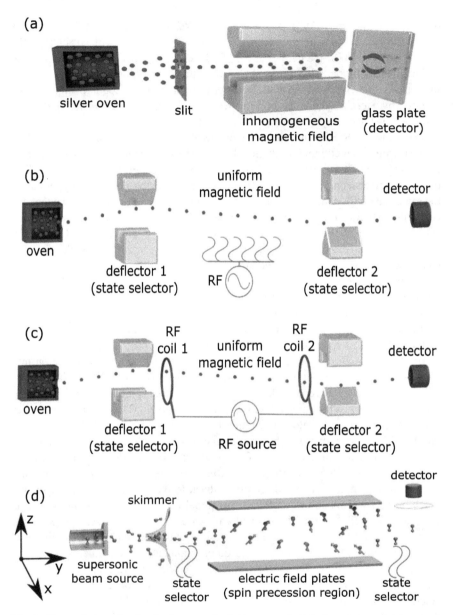

Fig. 2 The evolution of molecular beam methods for measuring dipole moments. (a) Stern and Gerlach's experiment for measuring magnetic moments and demonstrating space quantization. (b) Rabi's improved method for measuring magnetic moments. Between the two deflectors there is a uniform magnetic field and an oscillating field which, when resonant, changes the spin state. (c) A further improvement by Ramsey separates the oscillating field into two short regions. (d) Method for determining the electron's electric dipole moment by measuring the spin precession in an applied electric field. In all its key aspects, the technique is the same as Ramsey's method

due to the permanent EDM is $F_E = -\nabla W_E = R d_e \nabla E \, m_F / |F|$, and deflects states of opposite m_F in opposite directions. Taking $d_e = 10^{-29}$ e cm, which is close to the current upper limit [35], a field gradient of 100 kV/cm^2, and a speed of 200 m/s, the deflection after propagating 1 m is about 10^{-19} m. Clearly, this is not a good way to measure d_e. Nevertheless, the subsequent development of molecular beam methods inspired by Stern's techniques became so sensitive that measurements of d_e soon became feasible.

In the late 1930s, Rabi introduced a new idea that greatly improved Stern's method of measuring magnetic moments [36]. Instead of using the magnetic deflector as a measuring device, he used a pair of them as state selectors, as illustrated in Fig. 2b. The deflectors are arranged such that molecules reach a detector provided they remain in the same state so that they have the same magnetic moment throughout. Between the two deflectors, Rabi produced a uniform magnetic field, B_0, so that neighbouring m_F states are separated by an energy $\hbar\omega_0 = g\mu B_0$. In this region, an rf field of frequency $\omega \approx \omega_0$ resonantly drives transitions from one m_F state to another. Molecules that change m_F are deflected by the second magnet and miss the detector, resulting in a dip in the detected signal at $\omega = \omega_0$. This measurement of the resonant frequency, together with a measurement of B_0, determines the magnetic moment. The precision of this measurement is proportional to the interaction time with the rf field, so it's desirable to make this as long as possible. In practice however, this time is limited by the difficulty of keeping the fields uniform enough.

By the 1950s, Ramsey had solved this problem by separating the rf region into two short sections driven by the same oscillator [37], as illustrated in Fig. 2c. The first deflecting magnet prepares the molecules in a chosen spin state, say spin up. The first rf region rotates the spin so that it is orthogonal to B_0. The spin then precesses in the uniform magnetic field with angular frequency ω_0 for a time T. When $\omega = \omega_0$ the rf oscillation is in phase with the spin precession, so the second rf region rotates the spin in the same direction as the first, producing the spin-down state which will miss the detector. If there is a frequency difference $\omega - \omega_0 = \pm\pi/T$, the extra half rotation means that the spin will be driven back to the spin-up state in the second rf region and will reach the detector. The signal at the detector oscillates as ω is scanned, allowing ω_0 to be determined with an uncertainty inversely proportional to the free precession time T. Ramsey's method has such high precision that it is suitable for measuring the tiny electric dipole moments of fundamental particles. All that is needed is to add to B_0 a uniform electric field E_0, and then measure the change in the precession frequency when the direction of E_0 is reversed. This change is proportional to the EDM. This was the method used by Smith, Purcell and Ramsey in their first measurement of the neutron EDM [38], and the one used for all subsequent measurements of particle electric dipole moments.

Figure 2d illustrates a typical electron EDM measurement that uses Ramsey's molecular beam method. A molecular beam passes through a state selector, which could be a magnetic or electric deflector but in modern experiments is often a laser beam that optically pumps molecules to the desired state. Next, a first region of oscillating field aligns the spin along x. This can be done using an rf field, or a laser field that drives a Raman transition or optical pumping process. The spin now precesses around z due to the combination of μ interacting with B_0 and d_e interacting with E_0. Finally the spin direction is measured, for example by mapping its direction to a pair of states that are easily distinguished spectroscopically. The change in the spin precession angle that correlates with the reversal of E_0 determines d_e.

4 Current Status and Future Directions of eEDM Experiments

In the Standard Model of particle physics, the eEDM is predicted to be $d_e \approx 10^{-38}$ e cm [39]. Theories that extend the Standard Model often introduce new CP-violating interactions, which are needed to explain the observed asymmetry between matter and antimatter in the universe [40], and these new interactions lead to much larger eEDM values. Thus, eEDM measurements can be excellent probes of these theories. Early measurements used heavy atoms and yielded results consistent with zero, eventually assigning an upper limit of $|d_e| < 1.6 \times 10^{-27}$ e cm [41, 42]. Although more difficult to produce and control, heavy polar molecules can be far more sensitive to the eEDM than atoms [43, 44]. The sensitivity is proportional to the degree of polarization and therefore to the electric-field-induced mixing of opposite parity states. In atoms, these are different electronic states whose spacing is typically ~ 1 eV, but in molecules they are neighbouring rotational states whose energy spacing is about four orders of magnitude smaller, or the opposite parity states of an Ω-doublet where the spacing is even smaller still. Because the levels are closely spaced, only a modest electric field is needed to fully polarize the molecule. In this case, it is common to define an effective electric field $E_{\text{eff}} = RE$ which saturates to a maximum value $E_{\text{eff}}^{\text{max}}$. The effective field is enormous for some species, and its maximum value is often easy to reach. For example, $E_{\text{eff}}^{\text{max}} \approx 26$ GV/cm for YbF and ≈ 78 GV/cm for ThO.

The enormous effective fields make eEDM experiments with molecules very attractive, and measurements have been made using beams of YbF [45, 46], beams of ThO [35, 47], a cell of PbO vapour [48], and trapped HfF$^+$ molecular ions [49]. The results of these measurements are all consistent with zero, and the best upper limit is currently $|d_e| < 1.1 \times 10^{-29}$ e cm at the 90% confidence level [35]. Remarkably, this experiment and ones like it test theories that extend the Standard Model at an energy scale similar to, and even exceeding, the maximum collision energy of the Large Hadron Collider.

Given the great significance of eEDM experiments, it is natural to consider how to make the next leap in sensitivity. The uncertainty in measuring d_e scales as $1/(T\sqrt{N})$ where T is the spin precession time and N is the number of molecules used in the measurement. In a molecular beam experiment of length L, where the spin precession region occupies most of the space and the beam has diverged sufficiently that it fills the detector, T is proportional to L but N falls as $1/L^2$ because of the divergence of the beam. Consequently, there is no benefit in increasing L. This can be circumvented by cooling the molecules to much lower temperatures. A beam that is cooled in the transverse directions is collimated and can travel for long distances without spreading, allowing T to increase without reducing N. Going further, molecules that are cold enough can be launched into a fountain [50] or stored in a trap [51], giving access to even longer spin precession times. These ideas require molecules cooled to μK temperatures. Such low temperatures can be reached either by associating ultracold atoms into molecules [52], or by direct laser cooling [53]. EDM experiments using laser-cooled YbF, BaF, YbOH and TlF are all currently being developed [54–57]. They will have unprecedented sensitivity and tremendously exciting potential for new discoveries in fundamental physics.

5   Laser Cooling of Molecules: Principles

5.1   Laser Cooling Scheme

Figure 3 illustrates the energy level structure of a typical diatomic molecule, showing the electronic, vibrational, rotational, and hyperfine structure and the notation used to label the levels. For molecules to be slowed, cooled and trapped by radiation pressure, they must scatter many photons from the light, typically 10^4 or more. This calls for a cooling scheme where an upper level decays to only a few lower levels, so that only a few transitions need to be addressed. The inset to Fig. 3 shows an example of such a scheme. The upper level is the lowest level of positive parity in the first electronically excited state, labelled here as A, $v = 0$, $R = 0$, $+$. Electric dipole transitions to the X state must change the parity and obey the selection rule $\Delta R = 0, \pm 1$, which means that only the $R = 1$ rotational state is accessible.[1] However, the molecule can decay to any vibrational state, since there is no selection rule dictating how v can change in an electronic transition. The branching ratio to each vibrational state is mainly given by the square of the overlap integral between the vibrational wavefunctions in the lower and upper electronic states, which is known as the Franck Condon factor. For molecules where the optically active electron is not involved in the bonding, the sets of vibrational wavefunctions for the two electronic states are very similar. In this case, the branching ratio is close to 1 for the $\Delta v = 0$ transition and diminishes rapidly with increasing Δv. These molecules are the ones best suited to laser cooling because only a few vibrational bands need to be addressed, requiring just a few lasers, as indicated in the figure. Hyperfine components of these transitions can usually be addressed by adding radio-frequency sidebands to each laser using acousto-optic or electro-optic modulators.

---

[1] Often, R is not a good quantum number because it is coupled strongly to other angular momenta in the molecule, such as the orbital angular momenta of the electrons. In this case, R may be replaced by the relevant coupled angular momentum and the same principles apply.

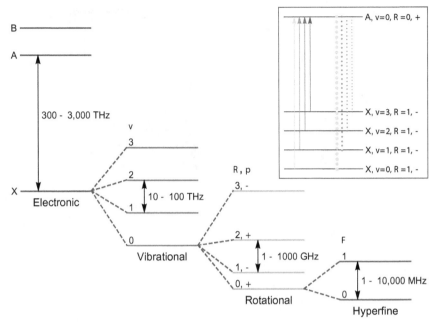

Fig. 3 Energy level structure of a typical diatomic molecule, with indicative transition frequencies. The ground electronic state is labelled X, and the excited states are A, B, etc. Transitions between electronic states are usually in, or near, the visible part of the electromagnetic spectrum. Each electronic state has a set of vibrational states, labelled by v. Vibrational transitions are usually in the mid-infrared. Each vibrational state has a set of rotational states, labelled by the rotational angular momentum, R, and the parity p. Rotational transitions are in the microwave regime. Each rotational state usually has a set of hyperfine states with total angular momentum F determined by the electronic angular momentum and the nuclear spin. In the example shown, both are 1/2. The inset shows the laser cooling scheme discussed in the text. Arrows show transitions driven by lasers, and the weights of the dotted lines indicate the relative branching ratios of the decay channels

5.2 Doppler and Sub-Doppler Cooling

Despite the complexity of the molecular structure outlined above, and the need to drive many transitions using several laser frequencies, the basic principles of laser cooling can be understood by focussing on just two or three levels. Figure 4a illustrates the principle of Doppler cooling applied in one dimension to a hypothetical two-level molecule. A molecule moving to the right with speed v interacts with a pair of identical, counter-propagating laser beams with wavevector k. The frequency of the light, ω, is slightly smaller than the molecular transition frequency, ω_0. The laser beam from the right is Doppler shifted closer to resonance, so the molecule scatters more photons from this beam and slows down as a result of this imbalanced radiation pressure. The force on the molecule due to each of the beams is

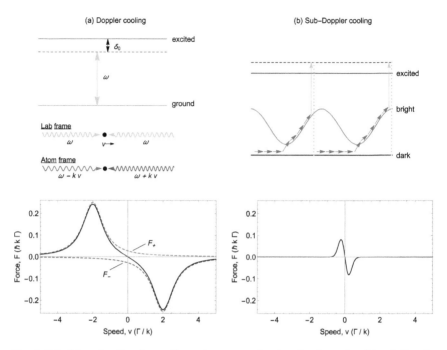

Fig. 4 Principles of **a** Doppler cooling, where a negative detuning is used and **b** sub-Doppler cooling, where a positive detuning is used

$$F_\pm = \pm \frac{\hbar k \Gamma}{2} \frac{I/I_s}{(1 + I/I_s + 4(\delta_0 \mp kv)^2/\Gamma^2)}, \tag{2}$$

where $\delta_0 = \omega - \omega_0$ is the detuning, Γ is the natural linewidth of the transition, I is the laser intensity and I_s is a characteristic intensity known as the saturation intensity. The graph in Fig. 4a shows these two forces as a function of v in the case where $I = I_s$ and $\delta_0 = -2\Gamma$. The solid line is their sum and shows that there is a force driving the molecule towards zero velocity. In addition to this cooling force, there is heating due to the randomly-directed momentum kicks associated with the photon scattering events. When the heating and cooling rates are balanced, the molecule reaches its equilibrium temperature. The minimum temperature for Doppler cooling is known as the Doppler limit and is $T_{D,min} = \hbar\Gamma/(2k_B)$.

Figure 4b illustrates a method of sub-Doppler cooling. Here, we distinguish two Zeeman sub-levels of the ground state. Due to the choice of states and the angular momentum selection rules, laser light of a given polarization cannot drive transitions from one of the ground states. We call this the dark state because it does not couple to the light. The other state is called the bright state. The bright state has an ac Stark shift which is positive when the detuning is positive. The dark state has no ac Stark shift because it does not couple to the light. If the two counter-propagating laser beams have different polarizations, neither parallel nor orthogonal, both the intensity and

Fig. 5 Typical form of the total force as a function of speed due to the combination of Doppler and sub-Doppler processes illustrated in Fig. 4. The two lines are for equal and opposite detunings: dashed red for negative, and solid blue for positive

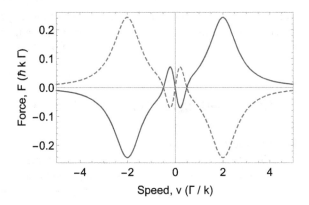

polarization of the light field will vary with position. This causes the ac Stark shift of the bright state to be modulated in position, setting up potential energy hills and valleys which the molecules move through. A molecule in the bright state will be excited by the light and optically pumped into the dark state. This is most likely to happen at positions where the intensity of the light field is high, which are also the positions where the ac Stark shift is largest, i.e. near the tops of the hills. A moving molecule can make a non-adiabatic transition back to the bright state because the polarization changes as it moves. This is most likely to happen where the energy gap between bright and dark states is smallest, i.e. near the bottom of the hills. As a result, molecules moving through the light field lose energy because they climb hills more often than they descend into valleys. The graph in Fig. 4b shows the typical force produced by this mechanism. It operates over a smaller range of velocities than Doppler cooling, and produces a smaller maximum force. Crucially however, the gradient near zero velocity, which is the damping constant, is substantially higher than for Doppler cooling. Furthermore, because the molecule spends much of its time in the dark state, there is less photon scattering, and thus a lower heating rate. Thus, for small velocities, the cooling rate is higher while the heating rate is lower, leading to much lower temperatures.

The Doppler cooling mechanism shown in Fig. 4a requires a negative detuning, while the sub-Doppler cooling mechanism requires a positive detuning.[2] The two mechanisms often appear together, resulting in the typical velocity-dependent force illustrated in Fig. 5. A negative detuning is useful for capturing molecules with a wide range of initial velocities and cooling them to lower velocity. However, the lowest temperatures are not reached because the total force has the wrong sign at low

---

[2]Note that there are other methods of sub-Doppler cooling, commonly used to cool atoms, that work for negative detunings. For molecules, it appears that sub-Doppler cooling always requires a positive detuning.

velocity. Once molecules are slow enough, the frequency of the light can be switched to a positive detuning so that the sub-Doppler mechanism cools them further. In this way, molecules have been cooled to temperatures far below the Doppler limit.

6   Laser Cooling of Molecules: Practice

Figure 6 illustrates an apparatus for laser cooling and trapping of molecules. The experiments begin with a molecular beam, a testament to the experimental power of the method developed by Stern and subsequent researchers. The cryogenic buffer gas sources described in Sect. 2 are ideal for this application because they deliver the critical combination of a high flux of molecules with a low initial speed. The illustration shows how this beam is laser cooled in the transverse directions, decelerated to low speed using radiation pressure, and then captured and cooled in a magneto-optical trap. We discuss each of these steps in turn.

6.1   *Transverse Laser Cooling of a Molecular Beam*

The density in a molecular beam drops with distance from the source because the beam spreads out as it propagates. Laser cooling can reduce the transverse temperature enormously, resulting in an intense, highly collimated molecular beam. The pioneering work on laser cooling of molecules was done at Yale [58]. They worked with a beam of SrF molecules and showed how to cool the beam in one transverse direction using both Doppler cooling and sub-Doppler cooling. Several other diatomic and polyatomic molecular species have since been cooled using similar

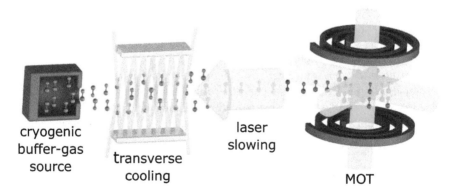

Fig. 6 An illustration of an apparatus for laser cooling and trapping of molecules. A beam of molecules from a cryogenic buffer gas source is cooled in the transverse directions, decelerated by the radiation pressure of a counter-propagating laser beam, and captured in a magneto-optical trap

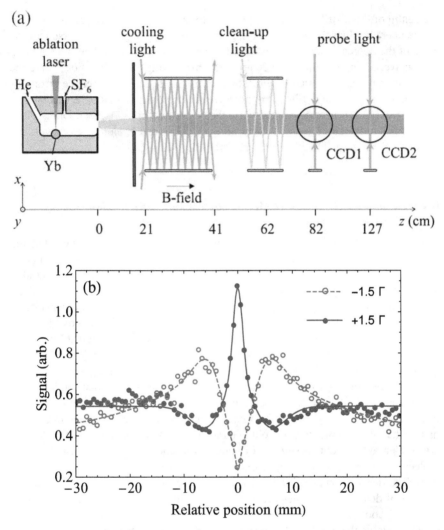

Fig. 7 Transverse laser cooling of YbF molecules. **a** Schematic of experiment. **b** Density distribution in one transverse direction following laser cooling in that direction. The lines are fits to a sum of four Gaussians. Adapted from [54]

methods [54, 59–64]. Figure 7a illustrates an experiment [54] to cool YbF molecules in one transverse direction, x. A beam of molecules from a cryogenic buffer gas source passes through a 20 cm long sheet of laser cooling light that forms a standing wave in the x direction. All molecules are then optically pumped to the lowest vibrational level by the clean-up light, and then detected by laser-induced fluorescence on a camera. Figure 7b shows the resulting density distribution of these molecules along x. When the detuning of the laser light is positive, a narrow peak appears at the centre of the distribution, corresponding to molecules that have been cooled to low temperature by sub-Doppler cooling. In this experiment, the transverse temperature of the cooled molecules was found to be below 100 μK. When the detuning is negative, a dip appears at the centre with broad wings on either side where the molecules have accumulated. The reason for this profile can be appreciated from the form of the force curve in Fig. 5. For negative detuning, the sub-Doppler mechanism forces molecules near zero velocity to a higher velocity, while Doppler cooling forces high velocity molecules towards lower velocities. As a result, molecules accumulate around the non-zero velocity where the force curve crosses zero.

6.2   Slowing a Molecular Beam with Radiation Pressure

Transverse laser cooling produces a highly collimated molecular beam, but the molecules still have a high forward speed. They can be decelerated using the radiation pressure of a laser beam propagating in the opposite direction. Here, it is essential to account for the changing Doppler shift as the molecules slow down. This can be done by chirping the frequency of the laser so that it follows the changing Doppler shift, or by broadening the frequency spectrum of the laser to cover the full range of Doppler shifts. Laser slowing of molecules was first demonstrated by the group at Yale using SrF [65], and similar methods have been applied to other molecules [66–69].

Figure 8 illustrates frequency-chirped laser slowing of a cryogenic buffer-gas beam of CaF molecules [69]. The black curves show the velocity distributions with no slowing applied, and the coloured curves show the distributions after slowing using various frequency chirps. The initial frequency of the laser is tuned to be resonant with molecules moving at about 180 m/s. When there is no chirp, the molecules are decelerated to about 100 m/s. They bunch up around this speed because the faster molecules are initially closer to resonance so are decelerated more than the slower ones. The distribution is shifted to lower velocities as the chirp increases, but the number of detected molecules drops at low velocities. This is because there is no transverse cooling in these experiments, so the beam diverges rapidly as it slows down, reducing the number of molecules that pass through the detector.

6.3   Trapping the Molecules

With the molecules slowed to low velocity, it becomes possible to trap them. Magneto-optical traps (MOTs) have been used to cool and trap atoms for decades [71]. In a MOT, counter-propagating laser beams result in a velocity-dependent force which cools the atoms, as described in Sect. 5.2. The detuning of the light is usually negative so that Doppler cooling, with its large capture velocity, is the dominant process. This alone does not trap the atoms because the force does not depend on position. To produce a position-dependent force, a magnetic field gradient is added,

Fig. 8 Radiation pressure slowing of a beam of CaF molecules. The laser light propagates in the opposite direction to the molecular beam and is frequency chirped. Black lines show the velocity distributions without slowing, and coloured lines are the distributions when the slowing is applied with various frequency chirps. The dashed lines show the resonant velocity at the beginning and end of the chirp. Adapted from [69]

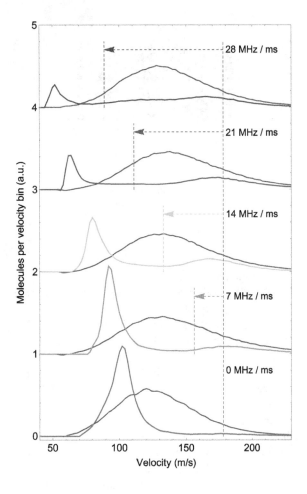

typically by using a pair of coils as illustrated in Fig. 6. The current flows in opposite directions in the two coils so that the magnitude of the field is zero at the centre and increases linearly in all directions away from this point. For a stationary atom at the field-zero, there is no net force because the atom is equally likely to scatter photons from any of the laser beams. When the atom is displaced, the Zeeman effect shifts the frequencies of transitions with $\Delta m_F = \pm 1$ in opposite directions, one closer to resonance and the other further away. These two transitions are driven by circularly polarized light of opposite handedness. By choosing the handedness of the beams in the correct way, the transition closest to resonance is always driven by the beam that pushes the atom back towards the field-zero. This traps the atoms.

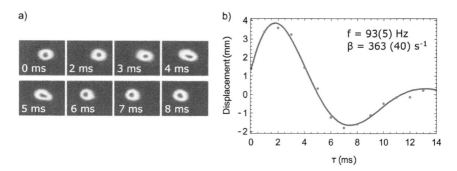

Fig. 9 Damped oscillations of CaF molecules in a magneto-optical trap. **a** Fluorescence images of the trapped molecules at various times after giving the molecules a radial push. **b** Red points: position of the centre of the cloud versus time. Blue line: fit to a damped harmonic oscillator model. Figure taken from [70]

The first three-dimensional magneto-optical trap of molecules was demonstrated in 2014 by the Yale group [72]. They captured a few hundred SrF molecules for about 50 ms at a density of 6×10^2 cm^{-3}, and cooled them to 2.3 mK. They went on to investigate several other trapping configurations [73–75] and were soon able to trap about 10^4 molecules for 500 ms at a density of 2.5×10^5 cm^{-3} and temperatures down to 250 μK. MOTs of CaF [70, 76, 77] and YO [78] molecules have also been produced, with steadily increasing number densities. Figure 9a is a sequence of pictures showing CaF molecules trapped in a MOT. Each picture is made by imaging the fluorescence of the trapped molecules onto a camera. The molecules were given a sudden push in the horizontal direction at time T=0 and the subsequent images show them oscillating in the trap. Figure 9b shows the displacement of the cloud versus T together with a fit to the motion of a damped harmonic oscillator. The results show that the MOT exhibits both a restoring force and a damping force, as expected.

Once molecules have been trapped in a MOT, they can be cooled to much lower temperature using the sub-Doppler cooling method described in Sect. 5.2. Typically, the magnetic field is turned off and the detuning of the lasers is switched from negative to positive. The molecules typically cool below the Doppler limit [76, 79, 80] in about 1 ms, and temperatures as low as 4 μK have been reached in this way [81–83]. At this point the cooling can be turned off and the molecules stored in conservative traps where their quantum states can be controlled and preserved for long periods. Ensembles of laser-cooled molecules have recently been confined in pure magnetic traps [80, 84] and in optical dipole traps [79], and single molecules have been held in tightly-focussed tweezer traps [85]. Coherent control of the hyperfine and rotational states of molecules has been studied [10, 84] and rotational superpositions with long coherence times have been demonstrated for trapped molecules [86].

7   Applications of Laser-Cooled Molecules

The ultracold molecules produced by direct laser cooling are well suited to a wide variety of exciting applications. Many of these applications require molecules trapped for long periods, long-lived coherences, and control over all degrees of freedom at the quantum level, often including the motional degree of freedom. Figure 10 illustrates four experimental approaches that satisfy some, or all, of these requirements. The molecules may be launched into a molecular fountain so that they are in free fall throughout a measurement [50, 87], or they could be trapped near the surface of a chip that integrates microscopic traps with superconducting microwave resonators [12]. Small, reconfigurable arrays of molecules can be produced using optical tweezer traps [85], while larger arrays can be made using optical lattices in one, two or three dimensions [88]. Here, we give an overview of future research directions using these platforms.

7.1   Testing Fundamental Physics

Ultracold molecules provide several avenues to constrain new theories or discover new physics [5]. As discussed in Sects. 3 and 4, the use of molecules to measure fundamental electric dipole moments is an amazingly powerful probe of symmetry-violating physics beyond the Standard Model. Other kinds of symmetry tests can also be done using molecules. For example, they can be used to explore parity violation in nuclei with unprecedented sensitivity [89]. Of particular interest is the measurement of nuclear anapole moments which arise from weak interactions within nuclei. A recent experiment with a beam of BaF has demonstrated the exceptional sensitivity achievable [90]. It is also of fundamental interest to measure the parity-violating energy difference between left- and right-handed chiral molecules, which is predicted but has not yet been observed [91]. The recent extension of laser cooling to polyatomic symmetric top molecules [63] shows that quite complex molecules can be cooled to sub-millikelvin temperature, making a parity-violation measurement using a laser cooled chiral species feasible in the future.

Precise measurements of molecular transition frequencies can also be used to test the idea that the fundamental constants may actually vary in time or space, or according to the local density of matter. Such variations are predicted by theories that aim to unify gravity with the other forces, and by some theories of dark energy [92, 93]. The frequencies of molecular transitions depend primarily on two fundamental constants, the fine-structure constant α and the proton-to-electron mass ratio $\mu = m_p/m_e$. The rotational and vibrational frequencies of molecules scale as μ^{-1} and $\mu^{-1/2}$ respectively, a direct dependence that an electronic transition in an atom does not have. Moreover, certain transitions have enhanced sensitivity to α or μ [94], sometimes because the transition energy results from a near cancellation between two large contributions of different origin. Astrophysical observations of atomic and molec-

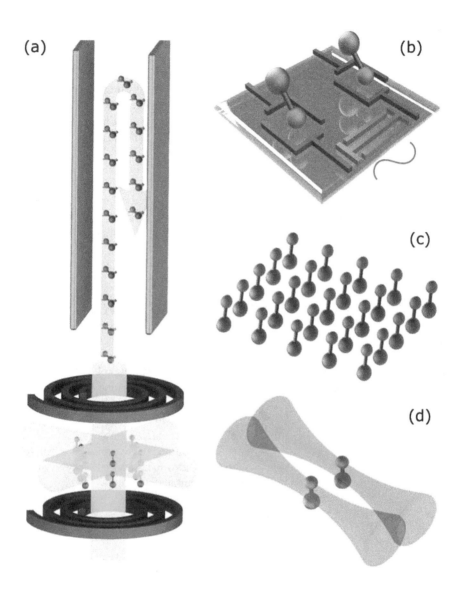

Fig. 10 Techniques for controlling ultracold molecules. **a** Molecules are launched into a fountain. **b** Molecules are stored on the surface of a chip using either electric or magnetic potentials created by planar electrodes or current-carrying wires. **c** Molecules are trapped in a 1D, 2D, or 3D lattice created by interfering counter-propagating lasers. **d** Molecules are trapped using an array of optical tweezers

ular spectra can be used to study variations on a cosmological timescale [5]. Here, laboratory measurements are important for establishing the present-day frequencies to high precision, as has been done using cold molecular beams of OH [95] and CH [96]. Atomic and molecular clocks can be used to set limits on present-day variations on a timescale of a few years. So far, the most stringent limits come from atomic clock measurements [97, 98], but molecular clocks are likely to contribute valuable information in the near future. For example, ultracold KRb molecules were recently used to set limits on the temporal variation of μ [99], a lattice clock of Sr_2 molecules is being developed [100], a molecular fountain of ultracold ammonia molecules has been demonstrated and could be used to search for variations in μ [87], and clocks based on the vibrational transitions of laser-cooled molecules look promising [101].

7.2 Collisions and Ultracold Chemistry

Molecules prepared at ultracold temperature in a single quantum state are ideal for studying how those molecules interact and what happens in a collision or chemical reaction [102]. With such a high degree of control, it becomes possible to explore how the rotational or hyperfine state influences the outcome of a collision, and to study collisions in a single partial wave regime. Electric or magnetic fields can be used to tune through collision resonances, and electric fields can be used to control the relative orientation of the colliding molecules. A fascinating recent advance in this direction is the study of collisions between individual laser-cooled molecules trapped in optical tweezers [103]. Two molecules, each prepared in a single quantum state, were brought together in a highly controlled way by merging the two separate tweezers, and the collisional loss rate measured for several choices of state. This experiment marks the first contribution of laser-cooled molecules to understanding ultracold chemistry. Some work in the ultracold regime has already been done using molecules assembled from ultracold atoms [52]. Examples include the control of chemical reactions though the choice of quantum state [104] or molecule orientation [105], and the controlled formation of a single molecule from a pair of atoms [106]. Direct laser cooling diversifies the set of ultracold molecules and molecular properties available for these studies, which is an exciting prospect for future research.

7.3 Quantum Simulation

It is important to understand the behaviour of systems consisting of many quantum particles all interacting with one another. These many-body quantum systems exhibit remarkable phenomena that are poorly understood at present, such as high temperature superconductivity, magnetism, the fractional Hall effect, and the structure of nuclei. We often use computer simulations to help understand complicated systems, but it is impossible for a (classical) computer to simulate more than a few tens of interacting quantum particles. Instead, we may try to engineer a well-controlled quantum system in such a way that it simulates a many-body problem that we wish to understand [107]. One system that has been developed with considerable success is an optical lattice of ultracold atoms. These lattices have been used to study important problems in condensed matter physics, such as the quantum phase transition between a superfluid and a Mott insulator [108], and models of antiferromagnetism [109]. However, the variety of systems that can be simulated is limited because atoms only have short-range interactions, meaning that they only interact appreciably when they are on the same site of the lattice. Complex many-body phenomena usually arise from long-range interactions among many particles, which atoms in a lattice struggle to emulate. A lattice of ultracold polar molecules solves this problem because the molecules interact through the long-range dipole-dipole interaction. This interaction has two main effects. First, the energy of the system depends on the configuration of the dipoles, and second, the interaction can mediate the transport of excitations from one site to another. Both effects have a long-range, anisotropic character, and can be controlled using a dc electric field or a microwave field resonant with a rotational transition. This makes for a tremendously rich environment for exploring the behaviour of interacting quantum systems [9]. Taking a first step in this direction, the effects of spin-exchange mediated by dipole-dipole interactions have been studied in a lattice of polar molecules [110]. Recent progress that will advance this field includes the formation of a Fermi degenerate gas of molecules [111], collisional cooling methods for molecules [112], the ever-increasing variety of polar molecules being brought into the ultracold regime, and the improvements in controlling their hyperfine and rotational states [10, 86, 113, 114]. It seems likely that, in the near future, lattices of ultracold polar molecules will significantly advance our understanding of strongly-interacting many-body quantum systems.

7.4 Quantum Information Processing

There are many proposals for using ultracold molecules for quantum information processing [11, 12, 115–117]. The hyperfine and rotational states of molecules have extremely long lifetimes and so can serve as stable qubits or qudits. By using microwave fields to drive rotational transitions, single qubit operations can be done rapidly and robustly using very mature microwave technology. The dipole-dipole interaction can be used to entangle pairs of molecules and perform two-qubit operations. Each molecule in an array can be addressed separately either by using a field gradient to shift the frequency of the qubit transition differently for each molecule, or by using an addressing laser to produce an ac Stark shift at a chosen site. One interesting approach for quantum information processing is an array of optical tweezer traps with a single molecule in each trap, as illustrated in Fig. 10(d) and recently demonstrated [85]. The molecules can be tightly confined and cooled to the motional ground state of the trap [118], and the array of molecules can be reconfigured as needed [119] in order to implement quantum gates between selected pairs. Another interesting approach is to trap the molecules near a surface using microscopic electric or magnetic traps, as illustrated in Fig. 10(b) and proposed in [12]. The same chip can support superconducting microwave resonators with small mode volumes and at frequencies that match the rotational frequency of the chosen molecule. By reaching the regime of strong coupling between trapped molecules and the resonator it becomes possible to transfer quantum information between a molecule and a microwave photon in the resonator, and to use the resonator to couple distant molecules to one another. This architecture is thus a hybrid quantum processor that combines the advantages of molecules for storing and processing quantum information with the advantages of photons for exchanging that information.

8 Summary

Over the past century, the humble molecular beam method has pushed forward the frontiers of knowledge in physics and chemistry. Today, molecules laser cooled to ultracold temperatures are an exciting and powerful platform for investigating the boundaries of modern scientific knowledge, including what might lie beyond the Standard Model of particle physics, how chemistry works at a fundamental level, and how quantum phenomena lead to emergent collective behaviors. This rich history and bright future has been strongly shaped by the visionary work of Otto Stern and his colleagues.

Acknowledgements We are grateful to Bretislav Friedrich and Horst Schmidt-Böcking for inviting us to contribute to this Festschrift, and to Luke Caldwell and Ed Hinds for their helpful feedback. The authors are supported by EPSRC (EP/P01058X/1), STFC (ST/S000011/1), the Royal Society, the John Templeton Foundation (grant 61104), the Gordon and Betty Moore Foundation (grant 8864), and the Alfred P. Sloan Foundation (grant G-2019-12505).

References

1. O. Stern, W. Gerlach, Z. Phys. **9**, 349 (1922)
2. O. Stern, Z. Phys **39**, 751 (1926)
3. R. Frisch, Z. Phys. **86**, 42 (1933)
4. D. DeMille, J.M. Doyle, A.O. Sushkov, Science **357**, 990 (2017)
5. M.S. Safronova, D. Budker, D. DeMille, D.F.J. Kimball, A. Derevianko, C.W. Clark, Rev. Mod. Phys. **90**, 025008 (2018)
6. D.S. Jin, J. Ye, Chem. Rev. **112**, 4801 (2012)
7. J. Toscano, H.J. Lewandowski, B.R. Heazlewood, Phys. Chem. Chem. Phys. **22**, 9180 (2020)
8. A. Micheli, G. Pupillo, H.P. Büchler, P. Zoller, Phys. Rev. A **76**, 043604 (2007)
9. A.V. Gorshkov, S.R. Manmana, G. Chen, J. Ye, E. Demler, M.D. Lukin, A.M. Rey, Phys. Rev. Lett. **107**, 115301 (2011)
10. J.A. Blackmore, L. Caldwell, P.D. Gregory, E.M. Bridge, R. Sawant, J. Aldegunde, J. Mur-Petit, D. Jaksch, J.M. Hutson, B.E. Sauer, M.R. Tarbutt, S.L. Cornish, Quantum. Sci. Technol. **4**, 014010 (2018)
11. D. DeMille, Phys. Rev. Lett. **88**, 067901 (2002)
12. A. André, D. DeMille, J.M. Doyle, M.D. Lukin, S.E. Maxwell, P. Rabl, R.J. Schoelkopf, P. Zoller, Nat. Phys. **2**, 636 (2006)
13. O. Stern, Z. Phys. **2**, 49 (1920)
14. O. Stern, Z. Phys. **3**, 417 (1921)

15. B.E. Sauer, J. Wang, E.A. Hinds, Phys. Rev. Lett. **74**, 1554 (1995)
16. M.R. Tarbutt, J.J. Hudson, B.E. Sauer, E.A. Hinds, V.A. Ryzhov, V.L. Ryabov, V.F. Ezhov, J. Phys. B **35**, 5013 (2002)
17. N.E. Bulleid, S.M. Skoff, R.J. Hendricks, B.E. Sauer, E.A. Hinds, M.R. Tarbutt, Phys. Chem. Chem. Phys. **15**(29), 12299 (2013)
18. G. Scoles (ed.), *Atomic and Molecular Beam Methods* (Oxford University Press, Oxford, 1988)
19. R. Campargue (ed.), *Atomic and Molecular Beams: The State of the Art 2000* (Springer, Berlin, 2001)
20. S.E. Maxwell, N. Brahms, R. deCarvalho, D.R. Glenn, J.S. Helton, S.V. Nguyen, D. Patterson, J. Petricka, D. DeMille, J.M. Doyle, Phys. Rev. Lett. **95**, 173201 (2005)
21. N.R. Hutzler, H.I. Lu, J.M. Doyle, Chem. Rev. **112**, 4803 (2012)
22. H.I. Lu, J. Rasmussen, M.J. Wright, D. Patterson, J.M. Doyle, Phys. Chem. Chem. Phys. **13**, 18986 (2011)
23. O.R. Frisch, O. Stern, Z. Phys. **85**, 4 (1933)
24. I. Estermann, O. Stern, Z. Phys. **86**, 132 (1933)
25. I. Estermann, O. Stern, Z. Phys. **86**, 135 (1933)
26. E.M. Purcell, N.F. Ramsey, Phys. Rev. **78**, 807 (1950)
27. H. Weyl, *Symmetry* (Princeton University Press, Princeton, New Jersey, 1952)
28. T.D. Lee, C.N. Yang, Phys. Rev. **104**, 254 (1956)
29. C.S. Wu, E. Ambler, R.W. Hayward, D.D. Hoppes, R.P. Hudson, Phys. Rev. **105**, 1413 (1957)
30. R.L. Garwin, L.M. Lederman, M. Weinrich, Phys. Rev. **105**, 1415 (1957)
31. J.I. Friedman, V.L. Telgedi, Phys. Rev **105**, 1681 (1957)
32. J.H. Christenson, J.W. Cronin, V.L. Fitch, R. Turlay, Phys. Rev. Lett. **13**, 138 (1964)
33. P.G.H. Sandars, Phys. Lett. **14**, 196 (1965)
34. H.S. Nataraj, B.K. Sahoo, B.P. Das, D. Mukherjee, Phys. Rev. Lett. **101**, 033002 (2008)
35. V. Andreev, D.G. Ang, D. DeMille, J.M. Doyle, G. Gabrielse, J. Haefner, N.R. Hutzler, Z. Lasner, C. Meisenhelder, B.R. O'Leary, C.D. Panda, A.D. West, E.P. West, X. Wu, Nature **562**, 355 (2018)
36. I.I. Rabi, S. Millman, P. Kusch, J.R. Zacharias, Phys. Rev. **55**, 526 (1939)
37. N.F. Ramsey, Phys. Rev. **78**, 695 (1950)
38. J.H. Smith, E.M. Purcell, N.F. Ramsey, Phys. Rev. **108**, 120 (1957)
39. M. Pospelov, A. Ritz, Phys. Rev. D **89**, 056006 (2014)
40. A.D. Sakharov, Phys.-Uspekhi **34**(5), 392 (1991)
41. S.A. Murthy, J.D. Krause, Z.L. Li, L.R. Hunter, Phys. Rev. Lett. **63**, 965 (1989)
42. B.C. Regan, E.D. Commins, C.J. Schmidt, D. DeMille, Phys. Rev. Lett. **88**, 071805 (2002)
43. P.G.H. Sandars, in *Atomic Physics 4*, ed. by G. zu Pulitz (Plenum, New York, 1975), p. 71
44. E.A. Hinds, Physica Scripta **T70**, 34 (1997)
45. J.J. Hudson, B.E. Sauer, M.R. Tarbutt, E.A. Hinds, Phys. Rev. Lett. **89**, 023003 (2002)
46. J.J. Hudson, D.M. Kara, I.J. Smallman, B.E. Sauer, M.R. Tarbutt, E.A. Hinds, Nature **473**, 493 (2011)
47. J. Baron, W.C. Campbell, D. DeMille, J.M. Doyle, G. Gabrielse, Y.V. Gurevich, P.W. Hess, N.R. Hutzler, E. Kirilov, I. Kozyrev, B.R. O'Leary, C.D. Panda, M.F. Parsons, E.S. Petrik, B. Spaun, A.C. Vutha, A.D. West, Science **343**, 269 (2014)
48. S. Eckel, P. Hamilton, E. Kirilov, H.W. Smith, D. DeMille, Phys. Rev. A **87**, 052130 (2013)
49. W.B. Cairncross, D.N. Gresh, M. Grau, K.C. Cossel, T.S. Roussy, Y. Ni, Y. Zhou, J. Ye, E.A. Cornell, Phys. Rev. Lett. **119**, 153001 (2017)
50. M.R. Tarbutt, B.E. Sauer, J.J. Hudson, E.A. Hinds, New J. Phys. **15**, 053034 (2013)
51. M.R. Tarbutt, J.J. Hudson, B.E. Sauer, E.A. Hinds, Faraday Discuss. **142**, 37 (2009)
52. K.K. Ni, S. Ospelkaus, M.H.G. de Miranda, A. Pe'er, B. Neyenhuis, J.J. Zirbel, S. Kotochigova, P.S. Julienne, D.S. Jin, J. Ye, Science **322**, 231 (2008)
53. M.R. Tarbutt, Contemp. Phys. **59**, 356 (2018)
54. J. Lim, J.R. Almond, M.A. Trigatzis, J.A. Devlin, N.J. Fitch, B.E. Sauer, M.R. Tarbutt, E.A. Hinds, Phys. Rev. Lett. **120**, 123201 (2018)
55. The NL-eEDM collaboration, P. Aggarwal, H.L. Bethlem, A. Borschevsky, M. Denis, K. Esa-

jas, P.A.B. Haase, Y. Hao, S. Hoekstra, K. Jungmann, T.B. Meijknecht, M.C. Mooij, R.G.E. Timmermans, W. Ubachs, L. Willmann, A. Zapara, Eur. Phys. J. D **72**, 197 (2018)
56. I. Kozyrev, N.R. Hutzler, Phys. Rev. Lett. **119**, 133002 (2017)
57. E.B. Norrgard, E.R. Edwards, D.J. McCarron, M.H. Steinecker, D. DeMille, S.S. Alam, S.K. Peck, N.S. Wadia, L.R. Hunter, Phys. Rev. A **95**, 062506 (2017)
58. E.S. Shuman, J.F. Barry, D. DeMille, Nature **467**, 820 (2010)
59. M.T. Hummon, M. Yeo, B.K. Stuhl, A.L. Collopy, Y. Xia, J. Ye, Phys. Rev. Lett. **110**, 143001 (2013)
60. I. Kozyrev, L. Baum, K. Matsuda, B.L. Augenbraun, L. Anderegg, A.P. Sedlack, J.M. Doyle, Phys. Rev. Lett. **118**, 173201 (2017)
61. B.L. Augenbraun, Z.D. Lasner, A. Frenett, H. Sawaoka, C. Miller, T.C. Steimle, J.M. Doyle, New J. Phys. **22**, 022003 (2020)
62. L. Baum, N.B. Vilas, C. Hallas, B.L. Augenbraun, S. Raval, D. Mitra, J.M. Doyle, Phys. Rev. Lett **124**, 133201 (2020)
63. D. Mitra, N.B. Vilas, C. Hallas, L. Anderegg, B.L. Augenbraun, L. Baum, C. Miller, S. Raval, J.M. Doyle, Science **369**, 1366 (2020)
64. R.L. McNally, I. Kozyrev, S. Vazquez-Carson, K. Wenz, T. Wang, T. Zelevinsky, New J. Phys. **22**, 083047 (2020)
65. J.F. Barry, E.S. Shuman, E.B. Norrgard, D. DeMille, Phys. Rev. Lett. **108**, 103002 (2012)
66. V. Zhelyazkova, A. Cournol, T.E. Wall, A. Matsushima, J.J. Hudson, E.A. Hinds, M.R. Tarbutt, B.E. Sauer, Phys. Rev. A **89**, 053416 (2014)
67. M. Yeo, M.T. Hummon, A.L. Collopy, B. Yan, B. Hemmerling, E. Chae, J.M. Doyle, J. Ye, Phys. Rev. Lett. **114**(22), 223003 (2015)
68. B. Hemmerling, E. Chae, A. Ravi, L. Anderegg, G.K. Drayna, N.R. Hutzler, A.L. Collopy, J. Ye, W. Ketterle, J.M. Doyle, J. Phys. B **49**, 174001 (2016)
69. S. Truppe, H.J. Williams, N.J. Fitch, M. Hambach, T.E. Wall, E.A. Hinds, B.E. Sauer, M.R. Tarbutt, New J. Phys. **19**, 022001 (2017)
70. H.J. Williams, S. Truppe, M. Hambach, L. Caldwell, N.J. Fitch, E.A. Hinds, B.E. Sauer, M.R. Tarbutt, New J. Phys. **19**, 113035 (2017)
71. E. Raab, M. Prentiss, A. Cable, S. Chu, D. Pritchard, Phys. Rev. Lett. **59**, 2631 (1987)
72. J.F. Barry, D.J. McCarron, E.B. Norrgard, M.H. Steinecker, D. DeMille, Nature **512**, 286 (2014)
73. D.J. McCarron, E.B. Norrgard, M.H. Steinecker, D. DeMille, New J. Phys. **17**, 035014 (2015)
74. E.B. Norrgard, D.J. McCarron, M.H. Steinecker, M.R. Tarbutt, D. DeMille, Phys. Rev. Lett. **116**, 063004 (2016)
75. M.H. Steinecker, D.J. McCarron, Y. Zhu, D. DeMille, Chem. Phys. Chem. **17**, 3664 (2016)
76. S. Truppe, H.J. Williams, M. Hambach, L. Caldwell, N.J. Fitch, E.A. Hinds, B.E. Sauer, M.R. Tarbutt, Nat. Phys. **13**, 1173 (2017)
77. L. Anderegg, B.L. Augenbraun, E. Chae, B. Hemmerling, N.R. Hutzler, A. Ravi, A. Collopy, J. Ye, W. Ketterle, J.M. Doyle, Phys. Rev. Lett. **119**, 103201 (2017)
78. A.L. Collopy, S. Ding, Y. Wu, I.A. Finneran, L. Anderegg, B.L. Augenbraun, J.M. Doyle, J. Ye, Phys. Rev. Lett. **121**, 213201 (2018)
79. L. Anderegg, B.L. Augenbraun, Y. Bao, S. Burchesky, L.W. Cheuk, W. Ketterle, J.M. Doyle, Nat. Phys. **14**, 890 (2018)
80. D.J. McCarron, M.H. Steinecker, Y. Zhu, D. DeMille, Phys. Rev. Lett. **121**, 013202 (2018)
81. L.W. Cheuk, L. Anderegg, B.L. Augenbraun, Y. Bao, S. Burchesky, W. Ketterle, J.M. Doyle, Phys. Rev. Lett. **121**, 083201 (2018)
82. L. Caldwell, J.A. Devlin, H.J. Williams, N.J. Fitch, E.A. Hinds, B.E. Sauer, M.R. Tarbutt, Phys. Rev. Lett. **123**, 033202 (2019)
83. S. Ding, Y. Wu, I.A. Finneran, J.J. Burau, J. Ye, Phys. Rev. X **10**, 021049 (2020)
84. H.J. Williams, L. Caldwell, N.J. Fitch, S. Truppe, J. Rodewald, E.A. Hinds, B.E. Sauer, M.R. Tarbutt, Phys. Rev. Lett. **120**, 163201 (2018)
85. L. Anderegg, L.W. Cheuk, Y. Bao, S. Burchesky, W. Ketterle, K.K. Ni, J.M. Doyle, Science **365**, 1156 (2019)
86. L. Caldwell, H.J. Williams, N.J. Fitch, J. Aldegunde, J.M. Hutson, B.E. Sauer, M.R. Tarbutt,

Phys. Rev. Lett. **124**, 063001 (2020)

87. C. Cheng, A.P.P. van der Poel, P. Jansen, M. Quintero-Pérez, T.E. Wall, W. Ubachs, H.L. Bethlem, Phys. Rev. Lett. **117**, 253201 (2016)

88. S.A. Moses, J.P. Covey, M.T. Miecnikowski, B. Yan, B. Gadway, J. Ye, D.S. Jin, Science **350**, 659 (2015)

89. S.B. Cahn, J. Ammon, E. Kirilov, Y.V. Gurevich, D. Murphree, R. Paolino, D.A. Rahmlow, M.G. Kozlov, D. DeMille, Phys. Rev. Lett. **112**, 163002 (2014)

90. E. Altuntaş, J. Ammon, S.B. Cahn, D. DeMille, Phys. Rev. Lett. **120**, 142501 (2018)

91. S. Tokunaga, C. Stoeffler, F. Auguste, A. Shelkovnikov, C. Daussy, A. Amy-Klein, C. Chardonnet, B. Darquié, Mol. Phys. **111**, 2363 (2013)

92. J.P. Uzan, Living Rev. Relativity **14**, 1 (2011)

93. K.A. Olive, M. Pospelov, Phys. Rev. D **77**, 043524 (2010)

94. C. Chin, V.V. Flambaum, M.G. Kozlov, New J. Phys. **11**, 055048 (2009)

95. E.R. Hudson, H.J. Lewandowski, B.C. Sawyer, J. Ye, Phys. Rev. Lett. **96**, 143004 (2006)

96. S. Truppe, R.J. Hendricks, S.K. Tokunaga, H.J. Lewandowski, M.G. Kozlov, C. Henkel, E.A. Hinds, M.R. Tarbutt, Nat. Commun. **4**, 2600 (2013)

97. N. Huntemann, B. Lipphardt, C. Tamm, V. Gerginov, S. Weyers, E. Peik, Phys. Rev. Lett. **113**, 210802 (2014)

98. R.M. Godun, P.B.R. Nisbet-Jones, J.M. Jones, S.A. King, L.A.M. Johnson, H.S. Margolis, K. Szymaniec, S.N. Lea, K. Bongs, P. Gill, Phys. Rev. Lett. **113**, 210801 (2014)

99. J. Kobayashi, A. Ogino, S. Inouye, Nat. Comms. **10**, 3771 (2019)

100. S.S. Kondov, C.H. Lee, K.H. Leung, C. Liedl, I. Majewska, R. Moszynski, T. Zelevinsky, Nat. Phys. **15**, 1118 (2019)

101. M. Kajita, J. Phys. Soc. Jpn. **87**, 104301 (2018)

102. J.L. Bohn, A.M. Rey, J. Ye, Science **357**, 1002 (2017)

103. L.W. Cheuk, L. Anderegg, Y. Bao, S. Burchesky, S. Yu, W. Ketterle, K.K. Ni, J.M. Doyle, Phys. Rev. Lett. **125**, 043401 (2020)

104. S. Ospelkaus, K.K. Ni, D. Wang, M.H.G. de Miranda, B. Neyenhuis, G. Quéméner, P.S. Julienne, J.L. Bohn, D.S. Jin, J. Ye, Science **327**, 853 (2010)

105. M.H.G. de Miranda, A. Chotia, B. Neyenhuis, D. Wang, G. Quéméner, S. Ospelkaus, J.L. Bohn, J. Ye, D.S. Jin, Nat. Phys. **7**, 502 (2011)

106. L.R. Liu, J.D. Hood, Y. Yu, J.T. Zhang, N.R. Hutzler, T. Rosenband, K.K. Ni, Science **360**, 900 (2018)

107. R.P. Feynman, Int. J. Theo. Phys. **21**, 467 (1982)

108. M. Greiner, O. Mandel, T. Esslinger, T.W. Hänsch, I. Bloch, Nature **415**, 39 (2002)

109. A. Mazurenko, C.S. Chiu, G. Ji, M.F. Parsons, M. Kanász-Nagy, R. Schmidt, F. Grusdt, E. Demler, D. Greif, M. Greiner, Nature **545**, 462 (2017)

110. B. Yan, S.A. Moses, B. Gadway, J.P. Covey, K.R.A. Hazzard, A.M. Rey, D.S. Jin, J. Ye, Nature **501**, 521 (2013)

111. L.D. Marco, G. Valtolina, K. Matsuda, W.G. Tobias, J.P. Covey, J. Ye, Science **363**, 853 (2019)

112. H. Son, J.J. Park, W. Ketterle, A.O. Jamison, Nature **580**, 197 (2020)

113. J.W. Park, Z.Z. Yan, H. Loh, S.A. Will, M.W. Zwierlein, Science **357**, 372 (2017)

114. F. Seeßelberg, X.Y. Luo, M. Li, R. Bause, S. Kotochigova, I. Bloch, C. Gohle, Phys. Rev. Lett. **121**, 253401 (2018)

115. S.F. Yelin, K. Kirby, R. Côté, Phys. Rev. A **74**, 050301 (2006)

116. K.K. Ni, T. Rosenband, D.D. Grimes, Chem. Sci. **9**, 6830 (2018)

117. R. Sawant, J.A. Blackmore, P.D. Gregory, J. Mur-Petit, D. Jaksch, J. Aldegunde, J.M. Hutson, M.R. Tarbutt, S.L. Cornish, New J. Phys. **22**, 013027 (2020)

118. L. Caldwell, M.R. Tarbutt, Phys. Rev. Res. **2**, 013251 (2020)

119. D. Barredo, S. de Léséleuc, V. Lienhard, T. Lahaye, A. Browaeys, Science **354**, 1021 (2016)

Permissions

All chapters in this book were first published by Springer; hereby published with permission under the Creative Commons Attribution License or equivalent. Every chapter published in this book has been scrutinized by our experts. Their significance has been extensively debated. The topics covered herein carry significant information for a comprehensive understanding. They may even be implemented as practical applications or may be referred to as a beginning point for further studies.

The contributors of this book come from diverse backgrounds, making this book a truly international effort. We would like to thank all the contributing authors for lending their expertise to make the book truly unique. They have played a crucial role in the development of this book. Without their invaluable contributions this book wouldn't have been possible. They have made vital efforts to compile up to date information on the varied aspects of this subject to make this book a valuable addition to the collection of many professionals and students.

This book was conceptualized with the vision of imparting up-to-date and integrated information in this field. To ensure the same, a matchless editorial board was set up. Every individual on the board went through rigorous rounds of assessment to prove their worth. After which they invested a large part of their time researching and compiling the most relevant data for our readers.

The editorial board has been involved in producing this book since its inception. They have spent rigorous hours researching and exploring the diverse topics which have resulted in the successful publishing of this book. They have passed on their knowledge of decades through this book. To expedite this challenging task, the publisher supported the team at every step. A small team of assistant editors was also appointed to further simplify the editing procedure and attain best results for the readers.

Apart from the editorial board, the designing team has also invested a significant amount of their time in understanding the subject and creating the most relevant covers. They scrutinized every image to scout for the most suitable representation of the subject and create an appropriate cover for the book.

The publishing team has been an ardent support to the editorial, designing and production team. Their endless efforts to recruit the best for this project, has resulted in the accomplishment of this book. They are a veteran in the field of academics and their pool of knowledge is as vast as their experience in printing. Their expertise and guidance has proved useful at every step. Their uncompromising quality standards have made this book an exceptional effort. Their encouragement from time to time has been an inspiration for everyone.

The publisher and the editorial board hope that this book will prove to be a valuable piece of knowledge for students, practitioners and scholars across the globe.

List of Contributors

Andres F. Ordonez
Max-Born-Institut, Max-Born-Str. 2A, 12489 Berlin, Germany

Olga Smirnova
Max-Born-Institut, Max-Born-Str. 2A, 12489 Berlin, Germany
Technische Universität Berlin, Straße des 17. Juni 135, 10623 Berlin, Germany

Gerard Meijer
Fritz-Haber-Institut der Max-Planck-Gesellschaft, Faradayweg 4-6, 14195 Berlin, Germany

S. Eckart, G. Gruber and T. Jahnke
Institut für Kernphysik, Universität Frankfurt, 60438 Frankfurt, Germany

H. Schmidt-Böcking,
Institut für Kernphysik, Universität Frankfurt, 60438 Frankfurt, Germany
Roentdek GmbH, 65779 Kelkheim, Germany

H. J. Lüdde
Institut für Theoretische Physik, Universität Frankfurt, 60438 Frankfurt, Germany

L. P. Schmidt, M. Schöffler and R. Dörner
Institut für Kernphysik, Universität Frankfurt, 60348 Frankfurt, Germany

V. Mergel
Patentconsult, 65052 Wiesbaden, Germany

O. Jagutzki, A. Czasch, K. Ullmann, S. Schößler and S. Voss
Roentdek GmbH, 65779 Kelkheim, Germany

R. Ali
Department of Physics, The University of Jordan, Amman 11942, Jordan

V. Frohne
Department of Physics, Holy Cross College, Notre Dame, IN 46556, USA

T. Weber and M. Prior
Chemical Sciences, LBNL, Berkeley, CA 94720, USA

A. Landers
Department of Physics, Auburn University, Auburn, AL 36849, USA

D. Fischer, M. Schulz and R. Olson
Department of Physics, Missouri S&T, Rolla, MO 65409, USA

A. Dorn and R. Moshammer
MPI für Kernphysik, 69117 Heidelberg, Germany

L. Spielberger
GTZ, 65760 Eschborn, Germany

J. Ullrich
PTB, 38116 Brunswick, Germany

C. L. Cocke
Department of Physics, Kansas State University, Manhattan, KS 66506, USA

Manfred Faubel
Max-Planck-Institut für Dynamik und Selbstorganisation, Göttingen, Germany

Mark Keil, Omer Amit, Or Dobkowski, Yonathan Japha, Samuel Moukouri, Daniel Rohrlich, Zina Binstock, Yaniv Bar-Haim, Menachem Givon, David Groswasser, Yigal Meir and Ron Folman
Department of Physics, Ben-Gurion University of the Negev, Be'er Sheva 84105, Israel

Shimon Machluf
Analytics Lab, Amsterdam, The Netherlands
Department of Physics, Ben-Gurion University of the Negev, Be'er Sheva 84105, Israel

Yair Margalit
Research Laboratory of Electronics, MIT-Harvard Center for Ultracold Atoms, Department of Physics, Massachusetts Institute of Technology, Cambridge, MA 02139, USA
Department of Physics, Ben-Gurion University of the Negev, Be'er Sheva 84105, Israel

Zhifan Zhou
Joint Quantum Institute, National Institute of Standards and Technology and the University of Maryland, College Park, Maryland 20742, USA
Department of Physics, Ben-Gurion University of the Negev, Be'er Sheva 84105, Israel

N. J. Fitch and M. R. Tarbutt
Centre for Cold Matter, Blackett Laboratory, Imperial College London, Prince Consort Road, London SW7 2AZ, UK

Index

A

Active Electron, 200

Alkali Atoms, 168

Angular Distributions, 93-94, 109, 117-118

B

Bulk Phase, 137, 143

C

Capacitor, 23

Charged Fragments, 34, 36, 39, 41, 98, 106

Chiral Molecules, 1-4, 11, 103, 209

Circular Dichroism, 2, 94, 118

Classical Perception, 53

Collapse Models, 184, 191

Condensation, 130, 186

Coupling Constants, 4, 6

D

Deceleration, 22-29, 31, 108

Diffraction Experiments, 158

Dipole Moments, 58, 192-193, 195, 197-198, 209

Doppler Cooling, 201-204, 206, 208

Doppler Limit, 202, 204, 208

E

Electric Deflection, 21, 133

Electric Dipole, 3, 5, 12, 58, 192-193, 195-198, 200, 209

Electric Field, 2-3, 20-21, 23, 25-27, 39-41, 63, 66, 73, 90-91, 106, 134, 196-199, 212

Electromagnetic Radiation, 21, 26

Electron Velocity, 41, 43, 48-50, 87

Electronic States, 199-201

Emission Directions, 92-93, 101

Emitted Electron, 44, 68, 88, 96, 100, 107

Enantiosensitivity, 3, 5, 7

F

Fragment Detection, 95

Fragmentation Process, 57, 103

Fundamental Physics, 192-193, 195, 200, 209

G

Grating Period, 158

Gravitational Field, 170

H

Harmonic Oscillator, 208

Hyperfine, 31, 200-201, 208, 211-212

I

Impact Parameters, 34, 38, 86, 88

Interference Structure, 47

M

Molecular Rotation, 7, 12, 93

Momentum Spectroscopy, 34, 50, 54, 56, 71, 78, 80, 111-112, 114, 116, 118

Multi-particle Coincidence, 36, 65, 109

N

Nuclear Scattering, 43, 71, 87-88, 106-107

O

Optical Detection, 160

Oscillations, 3, 123, 161, 166, 172, 184, 208

P

Permanent Dipole, 1, 4-8, 11-12

Photoelectron, 2, 93-94, 96-97, 99, 101, 108-110, 117-118, 121, 132-135, 137-153

Photoelectron Emission, 94, 132, 135, 137, 144-145, 151

Photoelectron Spectra, 133-135, 138, 140-145, 147-149, 151

Photoelectron Spectroscopy, 121, 132-133, 138-139, 142, 144, 150-153

Photoionization, 49, 83, 92-93, 109, 112, 115-116, 118-119, 145, 153

Photon Energy, 3, 49, 83, 98, 100, 108, 132-133, 136, 138, 140, 146-147, 149-151

Polarization Plane, 1-3, 6-7, 11

Precision Limits, 33-34, 56, 111

Projectile Energy, 60, 78, 80

Propagation Direction, 109, 128, 159

Q

Quantum Effects, 191

Quantum Gravity, 187, 191

Quantum Information, 193, 212-213

Quantum Interference, 189, 191

Quantum Measurement, 33-39, 48, 52-53, 56, 107, 111, 191

Quantum Mechanics, 19-20, 53, 154, 170, 173, 184, 190-191

Quantum Particle, 33

Quantum Physics, 173, 185

Quantum Reflection, 188

Quantum State, 20, 25, 44, 101, 193, 211

Quantum Superposition, 188, 191

Quasi-molecular Orbitals, 42, 44, 84, 105-106

R

Radiation Pressure, 200-201, 204, 206-207

Reflection Methods, 105-106

Rotational States, 199, 201, 208, 212

S

Slit Width, 47, 58

Solvation, 135-136

Space Quantization, 21, 36, 44, 105, 192, 195-197

Spatial Superposition, 170, 184

Spin Population, 157, 166, 178, 184

Spin Precession, 197-200

Stark-decelerator, 19-20, 26-27, 29-30

Sub-doppler Cooling, 201-204, 206, 208

Substantial Fraction, 195

Super-sonic Beam, 74, 78

Synchrotron Radiation, 114, 133, 138, 145, 152-153

T

Thermal Beams, 165

Thermal Motion, 78, 194

Transverse Extraction, 73, 76-77

Transverse Laser Cooling, 204-206

U

Ultracold Molecules, 192-193, 209-212

Ultrahigh Vacuum, 122-123, 128, 133

Uniaxial Orientation, 1, 7, 11

Unprecedented Sensitivity, 200, 209

V

Vacuum System, 77, 122

W

Wave Vector, 2

Wavepacket, 155-156, 162, 165-166, 168, 171

Printed in the USA
CPSIA information can be obtained
at www.ICGtesting.com
JSHW011643301023
51110JS00004B/14